The Technical Communicator's HANDBOOK OF TECHNOLOGY TRANSFER

by

Hyman Olken

OLKEN PUBLICATIONS

2830 Kennedy Street, Livermore, CA 94550

U.S.A.

ACKNOWLEDGMENTS

My thanks and gratitude to the several friends whose help, in ways big or small, helped to make this book possible.

Sources from which text excerpts or illustrations were taken are indicated in the references at the end of each chapter.

ISBN 0-934818-01-0

PATENTS

Patent considerations are naturally involved in all the technology developments discussed in this book. However, they constitute a professional field in themselves, so are omitted from discussion in this text. The reader interested in the patent issues related to technology transfer is advised to consult an excellent review of the whole area of patents which was published in the Special Issue on Patents, IEEE TRANSACTIONS ON PROFESSIONAL COMMUNICATION, June 1979, Vol. PC-22, No. 2. The comprehensive coverage on patent subjects in this issue is indicated in its table of contents, which follows.

SPECIAL ISSUE ON PATENTS

R. J. Joenk	**46**	Patents: Incentive to Innovate *and* Communicate—An Introduction
H. Skolnik	**59**	Historical Aspects of Patent Systems
B. M. Vanderbilt	**63**	The Plight of the Independent Inventor
H. J. Nussbaumer	**68**	Patents and the Engineer
W. G. Wolber	**73**	The Business Value of Patents
G. R. White	**77**	Management Criteria for Effective Innovation
R. V. Hughson	**83**	The Right Way to Keep Laboratory Notebooks
W. L. Franz and J. S. Child, Jr.	**86**	Good Habits Before Filing a Patent Application
C. M. Wright	**89**	Publication, Public Use, and Sale as Bars to Patenting
G. A. Hauptman	**92**	Patenting Inventions Based on Algorithms
K. J. Dood	**95**	The U.S. Patent Classification System
P. J. Terragno	**101**	Patents as Technical Literature
J. F. Cottone	**105**	Writing an Invention Disclosure
	109	Guide for Patent Drawings
J. T. Maynard	**112**	How to Read a Patent
A. B. Kimball, Jr.	**119**	Patenting U.S. Inventions Abroad
T. M. Noone	**123**	Trade Secret vs Patent Protection
	128	Information for Authors and Readers

ACRONYMS

AEC	Atomic Energy Commission
ANL	Argonne National Laboratory
APL	Applied Physics Laboratory (Johns Hopkins University)
BNL	Brookhaven National Laboratory
DoD	Department of Defense
DoE	Department of Energy
ERDA	Energy Research and Development Administration (Predecessor to DoE)
JPL	Jet Propulsion Laboratory
LASL	Los Alamos Scientific Laboratory
LBL	Lawrence Berkeley Laboratory
LLL	Lawrence Livermore Laboratory (Now Lawrence Livermore National Laboratory)
NASA	National Aeronautics and Space Administration
NBS	National Bureau of Standards
NSF	National Science Foundation
SLAC	Stanford Linear Accelerator Center

PREFACE

Technology transfer--the commercial application, in private industry, of the technological developments created in Government R&D programs--is nowadays a vital function of Government R&D agencies, for two reasons.

1. American industry is decreasing its expenditures for basic research, so must resort increasingly to the stockpile of technology, created in Government R&D programs, as the source of new products and processes.

2. American industry is under intense pressure to meet technological competition from other industrial countries.

The members of the technical writer/editor (technical communicator) profession, employed in large numbers in Government R&D agencies, play a key role in promoting technology transfer in those agencies. The author had sixteen years of experience performing that role in one of our great national laboratories. This book is a condensation of that experience, to provide guidelines that other writer/editors can use to further the technology transfer function in the R&D agencies in which they are employed.

Hyman Olken
Livermore, CA
January 1980

CONTENTS

PART I - INCREASING THE FLOW OF SPIN-OFFS

		Page
	Preface	V
Ch. I.	The Overall Picture	1
Ch. II.	Help the Engineer Get His Story Across	5
Ch. III.	Dig Out and Publicize the Neglected Technology	19
Ch. IV.	Give a Historical Perspective on the Whole Technology	29

PART II - FINDING THE TECHNOLOGY TRANSFER INFORMATION

Ch. V.	Provide a Guide Through the Maze in Your Own Government Agency	43
Ch. VI.	Guide Through the R&D Center's Organization	47
Ch. VII.	Guide Through the R&D Center's R&D Programs	55
Ch. VIII.	Guide Through the R&D Center's Technical Information Activities	65
Ch. IX.	Case History - Guide Through National Bureau of Standards	95

PART I

INCREASING THE FLOW OF SPIN-OFFS

Ch. I

THE OVERALL PICTURE

Outline

		Page
1.	Help the Engineer Get His Story Across	1
2.	Dig Out and Publicize the Neglected Technology	1
3.	Give a Historical Perspective on the Whole Technology	2
4.	Provide A Guide Through the Maze in Your Own Government Agency	3

Ch. I

THE OVERALL PICTURE

Thousands of technical writer/editors (technical communicators) are employed in Government R&D agencies. Their primary function is to perform the routine writing tasks of those agencies such as preparing manuals, R&D project progress reports, specifications, proposals, etc. In addition they are often assigned to help the engineer, who made a technological development, to prepare an article or paper on it. In this capacity the technical communicators are in a key position to effect large scale transfer of the technology, developed in Government R&D programs, to commercial application in private industry.

To fulfill that role they must carry out these four measures:

1. Help the Engineer Get His Story Across.

The technological development produced in a Government R&D center is lost to private industry if it does not become widely known and understood. That is, if an article or paper explaining its technological significance is not prepared and published in technical magazines or in conference proceedings.

The engineer who made the technological development is routinely helped, by the technical writer/editor assigned to his group, to produce an article or paper that is well written. But the competition to get material published in the technical field is so intense that having a well-written article or paper is not enough. In addition, the article or paper must have two "extras," which are:

(1) An opening section that strongly attracts reader interest.

(2) An illustration treatment that parallels and supports the text.

How the technical communicator can help the engineer achieve these two "extras" is detailed in Chapter II. The procedure is illustrated there by a few actual case histories.

2. Dig Out and Publicize the Neglected Technology.

Many valuable technology developments produced in Government R&D programs are lost to industry because they are never written up, or the write-ups are never published in the literature read by engineers and technologists. This happens because, in the rush to produce technological developments to fulfill contract schedules, the job of writing them up and getting the write-ups published gets lost in the shuffle.

These neglected technological developments must be dug up by the technical communicator working with the engineers who made them. He must then produce the write-ups and get them published himself, if he cannot get the engineers who made the developments to do it.

How such neglect and loss of valuable technology happens and what can be done about it are presented in Chapter III. To make the discussion there practical it is illustrated by a few actual case histories.

3. Give A Historical Perspective On the Whole Technology.

Sometimes industry will not be interested in the transfer of a particular technological development, created in a Government R&D program, even if that development is widely known and understood. To get interested, industry must be able to see good growth possibilities for a whole new technology of which that one development is a part. These possibilities are best conveyed by giving a historical perspective on that whole technology.

The technical communicator will occasionally find it possible to give such a "historical perspective" on a whole technology by merely reviewing the history of work done on that technology in the R&D center or agency in which he is employed.

For example, in the R&D program for developing high power glass lasers, the laser beam is produced by first generating a low power laser pulse of the desired time duration and pulse shape in a precision oscillator. This laser pulse is then built up to enormous energy levels by passing it through a chain of laser amplifiers.

As the laser beam crosses from one amplifier stage to the next, any minute misalignment between the amplifier stages causes some of the beam energy to be reflected back, and this backward-reflected energy can damage or destroy the glass optic elements of the laser system.

Thus, as a result of the work at LLL (Lawrence Livermore Laboratory) on the high energy laser program, a number of individual technology developments have been created to help overcome this internal self-damage of the laser optics--beam polarizers, beam rotators, beam cross-section shapers, damage-resistant glasses for laser optics, etc. These individual developments together constitute the bulk of the technology of overcoming internal damage to the optic elements of ultra-high energy glass lasers systems. A good historical perspective on this whole technology could therefore easily be prepared by reviewing the history of individual developments for this technology created at this one R&D center.*

Such "whole technologies" have been developed, over the years, in many R&D centers, and a "historical perspective" could be developed on each one by reviewing the evolution of that whole technology at the R&D center where it occurred. The writer/editors in those R&D centers are thus in an excellent position to write up such whole-technology historical perspectives. By doing so they can greatly stimulate the transfer to industry of many individual technological developments which, by themselves, are not of interest to industry, but when treated together as a whole technology, do present growth prospects that industry cannot resist.

The technique of preparing such "historical perspectives" is presented in Chapter IV and is illustrated there by a group of actual case histories.

4. Provide A Guide Through the Maze In Your Own Government Agency.

After the outsider has picked the Government-produced technology development he wants to transfer, he will quickly find that the article or conference paper from which he learned about it is not enough. To adapt that particular technological development to his company's commercial needs he will need the answers to many technical questions: Which materials to use in construction? How to scale it up or down in size? How to increase, reduce, or transform certain operating characteristics? Etc. He will therefore soon come knocking on the door of the R&D center in which the technological development he selected was created and seek information to answer all these questions.

---

*Many laboratories are engaged in a great variety of laser developments, so are all more or less engaged in solving the laser self-destruction problem. But the enormous power of the LLL lasers makes this problem most acute there, so the laboratory has concentrated heavily on its solution. See LASER PROGRAM ANNUAL REPORT - 1974. LLL Report UCRL-50021-74. Available from NTIS.

A great deal of the kind of information he desires is naturally available at the R&D center where the technological development to be transferred was created. But Government R&D agencies are not set up to make the obtaining of the information needed for technology transfer quick, direct, and easy. They are large organizations, with hundreds or even thousands of employees. They are organized in the complex organizational structures needed to carry out their R&D programs. The technical information they produce and its storage and retrieval are all organized to facilitate research and development, not technology transfer. For the outsider to find the information he needs to effect transfer of a technological development, in the R&D center that created it, is therefore like finding your way through a giant, complex maze.

However, the technical writer/editors of a Government R&D center can produce a guide that will enable the outsider to find his way successfully through this maze. Some Government agencies have already produced such guides and they have proven to be effective. The National Bureau of Standards is one. The greatest service which the technical writer/editors of a particular Government R&D center can do to improve the technology transfer process is therefore to ensure that their agency has such a guide, or to produce one if it does not.

The general procedure for preparing such a guide through the maze in one's own agency is presented in Chapter V. This guide through the maze in one's own R&D center is basically a composite of three separate guides--one through the complexities of an R&D center's organization; another through the R&D center's accumulation of R&D programs; and a third through the maze of the R&D center's technical information activities. These are treated separately in Chapters VI, VII and VIII, respectively.

These separate pictures of R&D center activities are combined in a single practical example of the overall guide for one major Government laboratory, which is presented in the last chapter, viz:

Ch. IX. Case History - Guide Through the National Bureau of Standards

NOTES

Ch. II

HELP THE ENGINEEER GET HIS STORY ACROSS

Outline

	Page
Opening Section that Arouses Reader Interest	5
Case History	5
Standard Practice	9
Illustration Treatment that Parallels and Supports the Text:	11
1. There Must Be Enough Illustrations	12
2. The Illustrations Must Be of Good Quality	12
3. The Illustration Treatment Must Be Organized in the Same Pattern as the Text Treatment	14

Ch. II

HELP THE ENGINEER GET HIS STORY ACROSS

The competition to get material published in the technical field is so intense that having a well-written article or paper is not enough. The article or paper must, in addition, have two "extras," which are:

1. An opening section that attracts reader interest the way honey draws flies.

2. An illustration treatment that parallels and supports the text.

Opening Section That Arouses Reader Interest

To attract reader interest, the article or paper must tell the reader at the outset (in the introduction or abstract) two things:

1. What's in it for him?

2. What's the key idea by which it is accomplished?

The reader leafing through a technical magazine or conference proceedings is subconsciously asking himself, as he looks at each article, "What's in it for me? What do I gain professionally that will make it worth investing enough of my time to read that article?" And the question is automatically followed by the next one: "What's the key idea by which it is accomplished?" A good example of this is provided by the following case history.

Case History

During the days when our scientists were exploring the radiation belts around the earth (about 1957) they sent rockets into outer space which carried radiation-measuring instruments. The data output of these instruments was sent to laboratories on the earth through telemetering channels.

Since the scientists did not know what levels of radiation would be encountered, they put onto each rocket a number of radiation-measuring instruments (count rate meters), each one able to measure a particular range of radiation intensity. To send data from each instrument to earth, each one was connected to a separate telemetry channel. This meant that each rocket would be equipped with a considerable number of telemetry channels. But telemetry channels are costly, so one engineer came up with this idea. Provide only two telemetry channels for each rocket. One channel would be used to identify the proper instrument selected for measuring the radiation intensity present in the radiation belt, the other channel would telemeter the output from that instrument to earth. He devised a special circuit that would automatically effect this arrangement. This development made possible huge savings in cost of the equipment because it reduced the number of telemetry channels required from a considerable number to only two.

When the engineer who made this development wrote an article on it, the writer/editor assigned to help him get it published directed him to add to the article an opening section that would tell the reader:

1. What was in it for him.

2. What is the key idea by which it was accomplished.

What the author wrote first is contained in Fig. 2-1. Here the reader has to grope through a whole page of detail, with no inkling of what's in it for him (the reader) until the last paragraph. Here you get an inkling that he had devised a scheme for automatic range changing in an instrument which has to cover many ranges. This is a valuable technical idea, but the reader would have lost interest long before he got to this point.

The author was advised to pull out of this garbled collection of ideas what it is that would make the device he has developed of value to other instrument designers. This he has done in the first sentence of the second paragraph in Fig. 2-2. The thing to be gained is the valuable economic result of reducing the number of telemetry channels from several to only two. The key idea by which this is accomplished is explained in the following (second) sentence of this paragraph. One now has an abstract which is easy to read and which immediately satisfies the reader's first interest on looking at an article - namely:

1. What's in it for him--i.e., how to reduce the number of telemetry channels, and thus the equipment cost.

2. The key idea by which this is accomplished--i.e., use one channel to identify the proper instrument selected for measuring the radiation, another channel to telemeter that instrument's data output to earth.

This is a good example of how poorly written an abstract

N-DECADE COUNT-RATE METER WITH AUTOMATIC

SCALE CHANGE FEATURE AND RESOLUTION OF ONE DECADE

ABSTRACT

The circuit described was designed for space flight application and is therefore transistorized and was laid out on two 4-inch square cards. Three decades were used but stacking the individual decade count rate circuits would permit coverage of any number of decades. Two channels of the telemetry system are needed; one for the count rate meter output and one for an indication of which count rate meter is being read out.

Pulses in our case from a BF_3 counter, after being shaped into quantized pulses are fed simultaneously into the inputs of all count rate meters being used. For a good part of the curve of output voltage vs. input rate, the output voltage is proportional to the log of the input rate. This portion of the curve is used when the response of the various CRM's is being selected such that the range to be covered is covered in only slightly overlapping stages. The output of each CRM can be gated off. After the gates the outputs are fed through "or" circuits and then all run into a summer circuit. All gates are open when the count rate is in the range of the first or lowest rate CRM, but no output is observed from any CRM except the first because of the "or" circuit since all other CRM have higher rate sensitivities. When the output of the first CRM reaches the end of the log portion of the curve, a special circuit closes the gate, disconnecting CRM number one from the system output. Now the second CRM can be read out. This process continues through n stages of CRM's.

A circuit senses which CRM's have reached the end of their range. This circuit provides telemetry information indicating which CRM is being read out.

Range changes are made in less than 1/2 ms which was compatible with CRM time constants. With minor modifications switching ranges could be done in a few μs.

The advantage of covering a large number of decades while still maintaining resolution of a single decade and automatically changing ranges are obvious to the fields of hazards control, chemistry and other areas.

Fig. 2-1. Initial draft of abstract is too long and confusing and omits content needed to arouse reader interest.

Abstract

The most efficient use of telemetry channels is always a paramount problem of space flight projects. Monitoring radiation fields in space often requires coverage of large ranges of count rate which, if count rate meters are to be used, means either a decrease in resolution due to coverage of the entire range with one meter or use of several telemetry channels if several different rate meters are used.

This paper describes a unique method permitting the desired resolution to be obtained by use of as many count rate meters as required while limiting the number of telemetry channels needed to two. A circuit switches one telemetry channel from one rate meter to the next as the input rate changes, while the other telemetry channel is used to monitor which rate meter is being read out.

Designing Avalanche Switching Circuits

High power can be obtained from transistors operated in the avalanche mode. Avalanche operation is reviewed and a criterion developed for selecting transistors with this property. The basic avalanche circuits are presented and factors influencing design and application discussed

Fig. 2-3

Example of abstract that arouses reader interest

Fig. 2-2 (Left)

Final abstract is short, well organized, and has content that arouses reader interest.

often is when first drafted, and what a good result can be achieved if one works at rewriting it until the function of the abstract--arousing reader interest--is achieved.

The paper was promptly accepted for the program of an IEEE conference and was published in the conference proceedings (1). The editor of a commercial electronics magazine, who read this paper, was so impressed with the idea of cost savings through reducing the number of telemetry channels that he had the engineer prepare a special article (2) on that circuit which automatically selected the appropriate instrument to measure the radiation intensity present. That article was also published. The engineer who made this development thus "got his story across"--fully.

A similar treatment solved the "reader interest" problem in the article (3) for publicizing another valuable technological development created at this laboratory (LLL)--the high-speed avalanche transistor circuit. Extremely fast circuits are needed in computers. Such fast circuit operation can be achieved by using transistors operating in the "avalanche" mode. So this information--what's in it for the reader (how to achieve ultra-fast circuits), and the key

August 1963 IEEE Proceedings

THE DETECTION OF CHIRPED RADAR SIGNALS BY MEANS OF ELECTRON SPIN ECHOES

W. B. Mims

Summary -- The application of the electron spin echo phenomenon to the detection of chirped microwave pulses is considered. The attainable time resolution is determined by the width of the paramagnetic resonance line, and the length of the received waveform is limited by the phase memory time of the spin packets in the active material. Pulse compressions of 1000:1 appear to be feasible. A typical device of this kind would require to be operated at liquid helium temperatures. It could be adapted so that in itself it provided maser amplification of the compound pulses.

Fig. 2-4

Typical author's abstract of technical paper. Fails to give cues as to what's in it for the reader and key idea by which it is accomplished.

August 1963 IEEE Proceedings

The Detection of Chirped Radar Signals by Means of Electron Spin Echoes (Mims, p. 1127) -- In a typical chirped radar system the transmitted pulse is subjected to a linear frequency sweep and the returned pulse is then passed through a frequency dispersive network in the receiver which delays one end of the pulse more than the other in such a way as to compress the pulse in time. This pulse compression technique makes it possible to transmit longer pulses with correspondingly more energy and yet maintain short pulse widths in the receiver for a high degree of radar resolution. In this paper the author shows that in the microwave region these frequency swept pulses can also be compressed--as much as 1000 to 1--by making use of the electron spin echo phenomenon in certain paramagnetic materials. The result is a highly original and important contribution to a practical radar problem.

Fig. 2-5

Commentary on the article abstracted in Fig. 2-4, which (1) explains what is a "chirped" radar; (2) why chirped radar is valuable to the reader; and (3) key idea by which "chirped" radar is accomplished.

idea by which it is accomplished (avalanche transistors)--are both stated in a block inserted at the front of the article (Fig. 2-3).*

Standard Practice

It is commonly assumed, in engineering and scientific fields, that the abstract which accompanies the article in

*Avalanche transistors have been bypassed by later technologies for achieving fast circuits, but this example well illustrates the point.

a journal or conference proceedings will do the job of arousing reader interest. Unfortunately, this is normally the exception rather than the rule. Consider this example.

The abstract presented in Fig. 2-4 is a typical one taken from a very respected technical journal--the IEEE PROCEEDINGS. It talks about "chirped" radar (4). But first the reader has to know, What is a "chirped" radar? Then he wants to know--What's the good of it? How will it help him achieve better radars? And finally: What's the key idea that makes chirped radars possible?

So the editor of this journal presents, in a feature called "Scanning the Issue," this "extra" section (5) that should have been added to the article to arouse reader interest. This opening section, reproduced in Fig. 2-5, makes it clear what a "chirped" radar is--one in which the pulses are shortened in length by the process of signal compression. It also tells "What's in it for the reader." "Chirping" the radar pulse makes possible more sensitive, more reliable radars. Finally, the key idea by which this is accomplished: The radar pulse is shortened by signal compression at its front and rear ends.

So commonly does this happen--that the abstract which accompanies the article does <u>not</u> arouse reader interest--that the editors of this prestigious journal have for decades included, in each issue of the IEEE PROCEEDINGS, an entire page called <u>Scanning the Issue</u>, in which a brief editorial commentary is given on each of the major articles in that issue, by which commentary the editor arouses reader interest in that article. This is done by telling, in that commentary:

1. What's in it for the reader.

2. What's the key idea by which it is accomplished.

The articles which are most often published in the IEEE PROCEEDINGS are lengthy ones which give a comprehensive picture of some important new development--like a new technology--that is of interest to almost the entire membership of IEEE. In those cases the authors' abstracts that accompany each article often very poorly serve the function of arousing reader interest. So the editors give such commentaries, on each major article, in which they arouse reader interest by the above defined technique.

A good example of this, taken from a more recent issue of the IEEE PROCEEDINGS (6), is presented in Fig. 2-6. In this figure the author's abstract that accompanies the

November 1977 IEEE Proceedings

Optical Methods of Electrical Measurement at High Voltage Levels

Robert E. Hebner, Jr., et al

Abstract - Optical methods to measure electric parameters and transmit the information from high voltage circuits to ground potential are described and evaluated in the light of the specific requirements of high-voltage measurement applications. The history and physics of a variety of optoelectrical methods found suitable for electrical measurement applications are introduced. Existing optical devices for measuring alternating, direct, and impulse currents and voltages in high-voltage circuits are reviewed with emphasis on the operation and features of several selected methods. The use of these techniques in industrial systems, in research laboratory apparatus, and in reference standards laboratories is discussed.

(A)

November 1977 IEEE Proceedings

SCANNING THE ISSUE

Optical Methods of Electrical Measurement at High Voltage Levels (Hebner, Malewski, and Cassidy, p. 1524) - The use of an optical link to transmit measurement signals from a high voltage circuit to sensitive instrumentation offers a number of advantages: the need for expensive and massive amounts of insulation is avoided, interference errors due to electrical coupling and ground currents are virtually eliminated, shielding requirements for the instruments are reduced, and so forth. Electrooptical techniques are particularly attractive for the measurement of high voltage pulses, such as those produced by lightning impulses, switching surges, and high-energy electrical discharge experiments. The measurement at high voltage levels in the steady state provides a fertile field for application of optical methods. Optical methods and devices also show promise for fruitful exploitation in standards work.

This invited paper's . . . emphasis is on measurements, at high voltage levels, of voltage, current, and electric fields using the Kerr, Faraday, and Pockels effects. The fundamental operating principles of the devices used are discussed at length and in a tutorial fashion . . . .

(B)

Fig. 2-6

Author's abstract (A) and "Scanning the Issue" commentary on the same article (B) which gives a picture of article's contents that arouses reader interest.

article is presented for comparison, to show how inadequately it usually serves to arouse in the reader a compelling interest to read that article.

Illustration Treatment That Parallels and Supports the Text

To convey a good picture of a technological development requires a symbiotic combination of text and illustrations. Therefore a technical article or paper should include, besides the text (verbal) treatment, an illustration treatment that parallels and supports the text. This means three things:

1. There must be enough illustrations. Each major idea covered in the text treatment should have a corresponding illustration.

For example, consider the article (7) to publicize a valuable technological development created at LLL--an electron beam welder. In this article on the welder the text first gives a general overall explanation of how the system works. This has a corresponding illustration in the functional block diagram (Fig. 2-7).

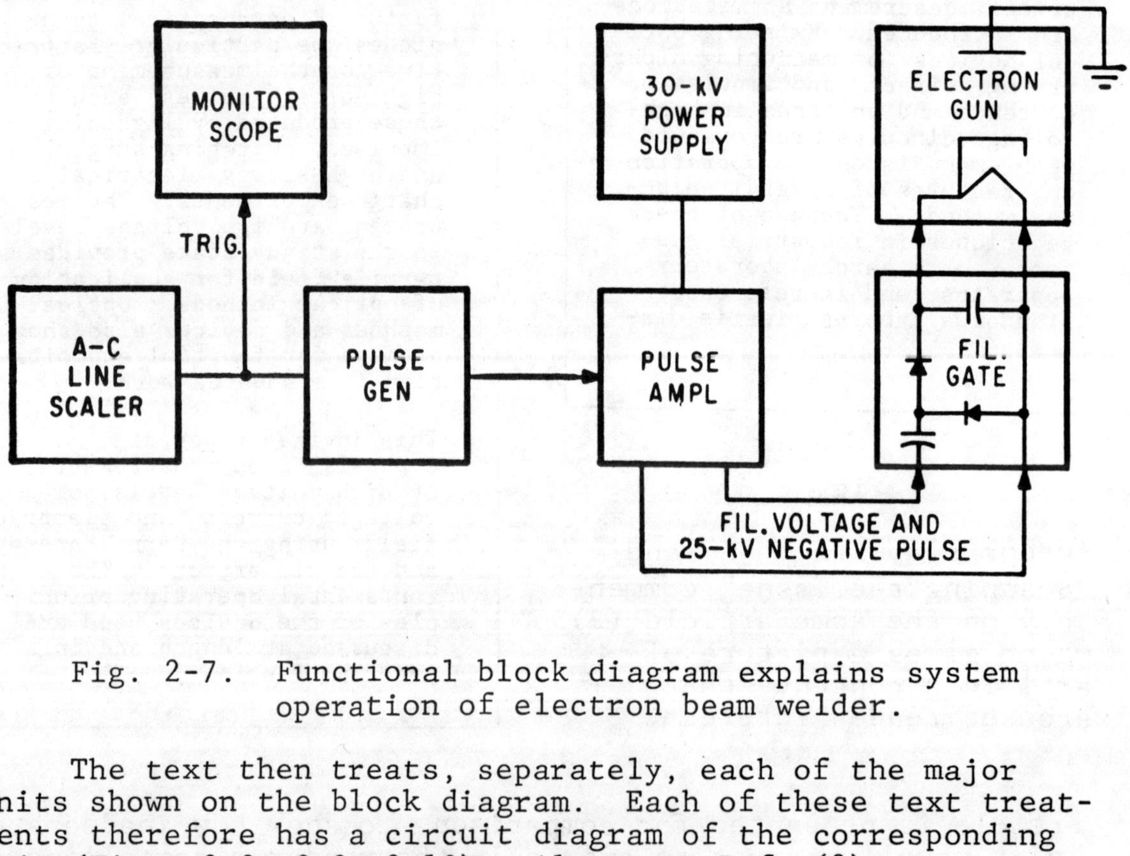

Fig. 2-7. Functional block diagram explains system operation of electron beam welder.

The text then treats, separately, each of the major units shown on the block diagram. Each of these text treatments therefore has a circuit diagram of the corresponding unit (Figs. 2-8, 2-9, 2-10). Also, see Ref. (8).

2. The illustrations must be of good quality. I.e., each illustration must support as well as parallel the text.

The basic unit of the text treatment is the paragraph or group of paragraphs. Each paragraph must therefore have the right content and be well organized.

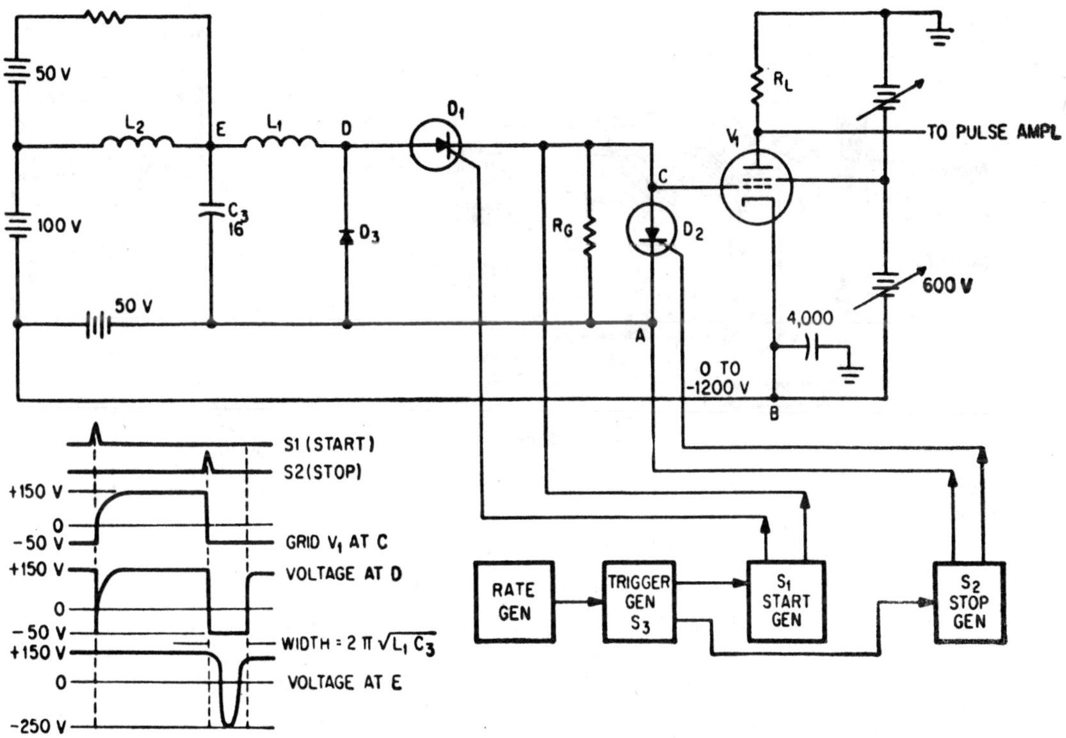

Fig. 2-8. Pulse generator circuit. Pulse wave forms show what happens in circuit. Last three wave forms are measured with respect to point B.

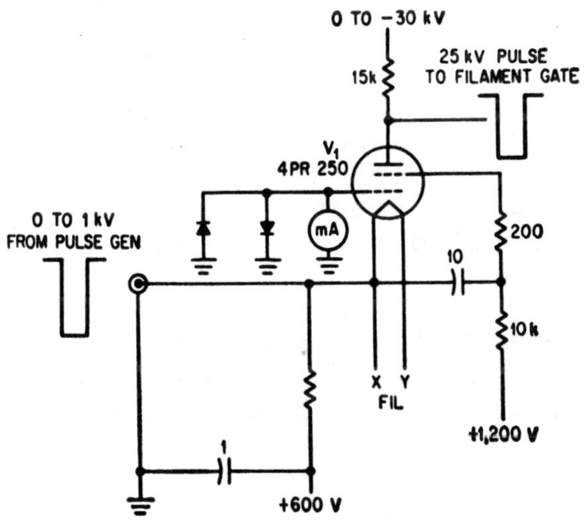

Fig. 2-9. Pulse amplifier circuit amplifies 1kV input pulse to 25 kV output pulse.

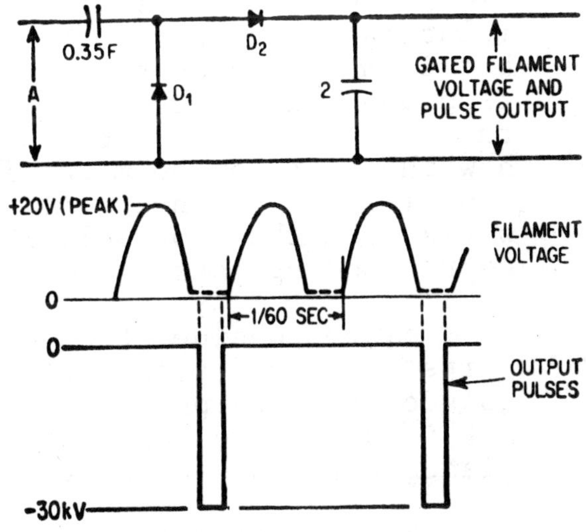

Fig. 2-10. Filament gate circuit. Output pulses selected during filament dead time.

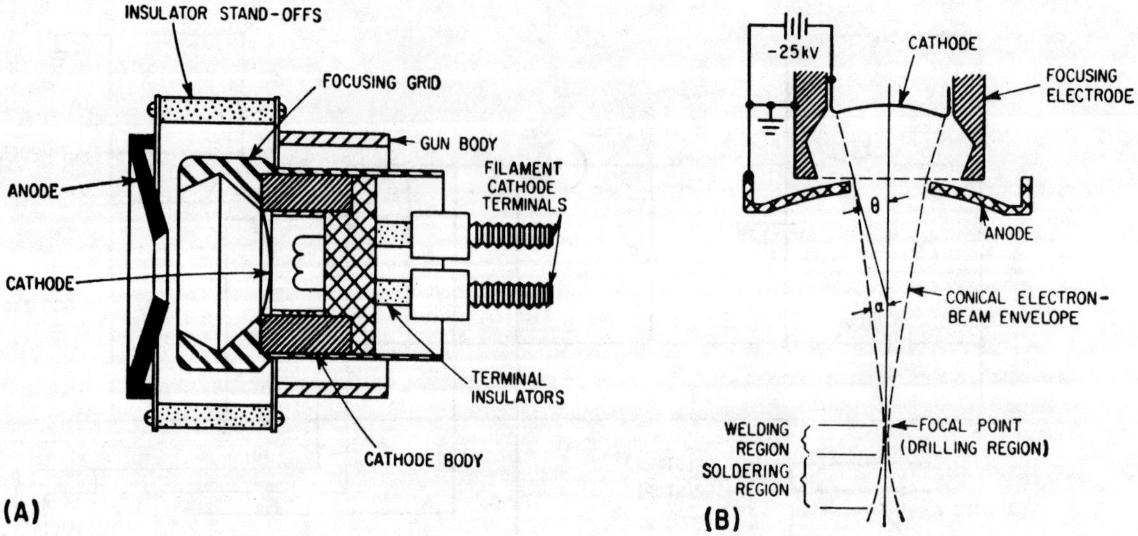

Fig.2-11. Electron gun (A) and electron beam envelope (B) showing the relationship of the gun electrodes required to obtain a sharp electrostatically focused beam.

The individual illustration is the basic unit of the illustration treatment. It must therefore have the right content and be well designed.

For example, the electron gun in this electron beam welder is a complex structure required to meet a combination of mechanical and electrical functions. The original illustration prepared for this was a cross section view, but it had no shading to indicate which parts were insulation and which were metallic conductors. Such shading was therefore added. Then a line diagram was added to show how the electron beam was formed. The resultant illustration (Fig.2-11) conveys at a glance how the electron gun works. It thus not only parallels but also supports the corresponding text.

Also, consider the circuit diagrams for pulsed circuits. If such a circuit diagram has no wave forms added to show what happens in the circuit, it is meaningless. But if such wave forms are added, it is apparent at a glance what goes on in the circuit (Figs. 2-8, 2-9, 2-10).

3. The illustration treatment must be organized in the same pattern as the text treatment.

For example, in this electron beam welder article, the text treatment gives first an overall picture of the whole system, then discusses separately each major unit shown on

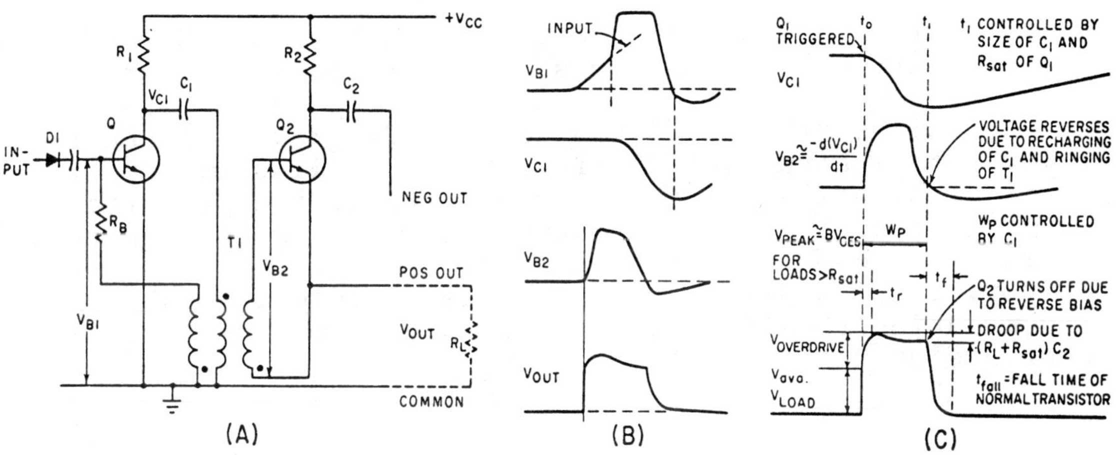

Fig.2-12. Complete circuit operation is communicated to reader's mind much faster by this diagram than by the corresponding word description. Diagram shows: Cascade avalanche circuit (A) will deliver positive or negative pulses; circuit wave forms (B): How circuit values control output wave form (C).

the block diagram, taken in order as one moves across the functional block diagram. The text treatment of each unit therefore has a corresponding circuit diagram which explains how that unit works (Figs.2-8,2-9,2-10). Thus the illustration treatment is organized in a pattern that closely follows the organization pattern of the text.

Another good example of all this is provided by the article (3) on avalanche transistor circuits, discussed above in connection with the provision of an opening section for the article or paper that attracts reader interest (Fig.2-3).

In that article the text treatment first presents an explanation of the avalanche transistor mechanism for producing very fast electrical circuit action. Then it presents, separately, the application of this principle to each of a number of different types of circuits.

The action inside a circuit is difficult to visualize. Therefore, although the text treatments of each of the different types of circuits is detailed and accurate, a circuit diagram has been added in each case to help the reader to visualize the circuit's operation. In one case the action inside the circuit is very complex, so a simple circuit diagram was not enough. Wave forms of the inputs and outputs were therefore added to the circuit diagram (Fig.2-12), thereby making the actions within the circuit easy to grasp. In short, an illustration was provided that supports as well as parallels the corresponding text.

Finally, by including a circuit diagram with the text treatment of each type of circuit, the illustration treatment was organized in the same pattern as the text treatment, thus creating a symbiotic relationship between them. In fact, so good was this symbiosis that according to the editor of the magazine in which the article was published, the magazine received and filled about 4,000 requests for reprints of that article!

REFERENCES

(1) N-Decade Count-Rate-Meter with Automatic Scale Change Feature and Resolution of One Decade. Stan Thomas, IEEE TRANSACTIONS ON NUCLEAR SCIENCE, Jan. 1963, pp. 36-41. Fig. 2-2 is from this paper. Copyright 1963, IEEE.

(2) N-Decade Count-Rate-Meter. Stan Thomas, ELECTRONIC DESIGN, Vol. 13, No. 6, Mar. 15, 1965, pp. 34-39.

(3) Designing Avalanche Switching Circuits. Richard P. Rufer, ELECTRONICS, April 7, 1961, pp. 83-87. Fig. 2-3 is from this article. Copyright 1961 by McGraw-Hill, Inc.

(4) The Detection of Chirped Radar Signals By Means of Electron Spin Echoes. W. B. Mims, PROCEEDINGS of IEEE, Vol. 51, p. 1127, Aug. 1963. Fig. 2-4 is from this paper. Copyright 1963, IEEE.

(5) Commentary on article in Ref. 4 in Scanning the Issue column, PROCEEDINGS OF IEEE, Vol. 51, p. 1082, Aug. 1963. Copyright 1963. IEEE. Fig. 2-5 is from this article.

(6) Optical Methods of Electrical Measurement at High Voltage Levels. Robert E. Hebner, Jr., Richard A. Malewski, and Esther C. Cassidy. PROCEEDINGS of IEEE, Vol. 65, No. 11, Nov. 1977, pp. 1524-1548. Fig. 2-6(A) is from this paper. Fig. 2-6(B) is from this issue, p. 1523. Copyright 1977, IEEE.

(7) Precision Electron Welder Uses Low Beam Voltage. Kristian Aaland and Ronald Hill, ELECTRONICS, Mar. 6, 1964. Copyright 1964 by McGraw-Hill, Inc.

(8) Figs. 2-7, 2-8, 2-9, 2-10, 2-11 are from Lawrence Livermore Laboratory Report UCID 4626-T, Electron Drilling and Welding Tool Using An Oxide Cathode, Kristian Aaland and Ronald Hill, Dec. 13, 1963. Available from National Technical Information Service, Springfield, VA 22161.

Consult also:

(9) Engineers CAN Write, Hyman Olken, AEROSPACE ENGINEERING, Vol. 21, No. 10, Oct. 1962, pp. 58, 59, 74, 75. This magazine is currently called ASTRONAUTICS and AEROSPACE ENGINEERING.

(10) Effective Technical Illustration. Hyman Olken. CA, COMMUNICATION ARTS, THE JOURNAL OF COMMERCIAL ART AND DESIGN, Jan. 1962, pp. 53-56. Address: 410 Sherman Ave., Palo Alto, CA 94303.

Ch. III

DIG OUT AND PUBLICIZE THE NEGLECTED TECHNOLOGY

Outline

	Page
The Lost Technology	19
"Lost Technology" Mines	19
Case Histories	21
Case I - Uninterruptable Power System	21
Case II - Remotely Controlled Seismic Network	23
Case III - Transportable Nuclear Explosives-Test Instrumentation System	25
Summary	26

Appendices

App. 3-1	Uninterruptable Power System	106
App. 3-2	Remotely Controlled Seismic Network	108
App. 3-3	Transportable Test Instrumentation System	110

Ch. III

DIG OUT AND PUBLICIZE THE NEGLECTED TECHNOLOGY

The Lost Technology

A good deal of the valuable technology that could be transferred is lost to private industry because adequate write-ups on those developments are not prepared and published. This occurs primarily for two reasons. Often, on completion of a technological development, the engineer who produced it is rushed off to another development job and is not given the time to write it up for publication. Or, sometimes he does not like to do the write-up job and deliberately evades it.

Another reason is that, in some cases, the development is produced by several groups working jointly. The job of writing it up for publication then "falls through the cracks" between them because no one group wants to carry the load for writing up the work done by the others. As a result, each group leaves it to the others to do the entire write-up of the development for publication. This usually ends with no write-up being produced at all.

Much of the technology thus lost can be recovered by the technical communicators in the Government R&D center digging out such neglected developments and writing them up for publication themselves, if they cannot get the engineers who produced the developments to do it.

How such technology loss happens and what can be done about it can be seen from these case histories given below.

Case I. Uninterruptable Power Supply System.

Case II. Remotely Controlled Seismic Network.

Case III. Transportable Instrumentation System for Nuclear Explosives Tests.

"Lost Technology" Mines.

An interesting pointer on where to find these neglected technological developments is the following. Government R&D centers operate mainly by fulfilling Government R&D contracts assigned to them. The completion of each contract requires the submission of an R&D report on the project that tells what was developed and how it was achieved. This is the case for the great majority of R&D contracts, which call for the conduct of specified research investigations designed to achieve particular research results. But a good deal of Government research funding is destined for a different type of contracts--those that call for the creation of expanded or advanced research facilities at Government R&D centers. In such cases the primary contractual requirement is satisfactory fulfillment of the specifications that define the research facilities to be created. Any write-up of the technology developed in the process of creating the new research facilities, and how that technology was achieved is a lesser, and may even not be a mandatory, contractual requirement.

In such cases the following occurs. If the new research facilities are being created for a laboratory which is competing with a few others in the same research program, as is, for instance, the case in the fusion research program, then the new technology developed in the course of creating the new research facility is abundantly written up and the write-ups are widely published in conference proceedings*, because such publication will build that laboratory's prestige in technical circles.

But if the laboratory for which the new research facilities are being created is not competing with other laboratories, then the technology developments produced in the course of creating its advanced research facilities receive only meager write-ups, as in Case III below; or no write-ups at all, as in Cases I and II below.

As a result, R&D programs for the creation of new research facilities make a good mine in which to find the nuggets of valuable new technology which were "lost" because they did not receive adequate write-up and publication.

---

*Thus, for example, the competition between laboratories in the fusion research program has produced a flood of published papers on the new facilities for laser fusion research at the Lawrence Livermore Laboratory; another string of papers has been published on the extremely complex torus fusion research reactor being built at Oak Ridge, and a similar flow of papers has been published on the high energy neutral particle beam facilities being developed at the Lawrence Berkeley Laboratory.

Case Histories

Case I. Uninterruptable Power System

The electrical power generation and distribution facilities in the U.S. are so advanced that it ordinarily never occurs to us that the steady flow of power from them might fail. But in recent years massive blackouts over huge parts of the country have occurred, and the possibility that interruptions in primary power supply can happen has become a reality which must be taken into account in the operation of equipment that is vulnerable to such power failures. In many cases this is taken care of by providing emergency power systems that are automatically turned on when the regular, utility-supplied power accidentally fails.

This problem was particularly acute in the radiochemistry building at the Lawrence Livermore Laboratory where huge, costly research machines like mass spectrometers are used that have large vacuum chambers kept at high vacuum by continuously running vacuum pumps. If electrical power to these pumps should fail, the vacuum chambers could be quickly contaminated by air leaking into them and would require a long, costly, clean-up operation to restore them to operating condition. Furthermore, the tests and experiments done with these machines may have to run continuously for as long as two years, so provision must be made to supply emergency power automatically if the utility-supplied power should fail.

Because of the difficult requirements for power supply at this building, commercially available equipment for emergency power supply would not be adequate. The requirements were that when utility power fails, emergency power should come on within <u>one cycle</u> of the 60-cycle power supply, or in 1/60 of a second. Furthermore, it was vital that during the change-over to emergency power, the voltage and frequency of the power should remain constant.*

To supply the heavy loads of emergency power (200 kw) required for this installation, it is common practice to use a system in which an ac electric motor, driven by the utility power line, drives an ac generator which furnishes the normal flow of power to the critical load. This motor-generator set has a heavy flywheel mounted on its shaft which stores up a considerable amount of energy. When the utility-line power to the ac motor fails, the stored energy in the flywheel is used to crank up a diesel engine which

---

*Equally stringent power requirements may also exist in other research facilities such as large computer systems and radars.

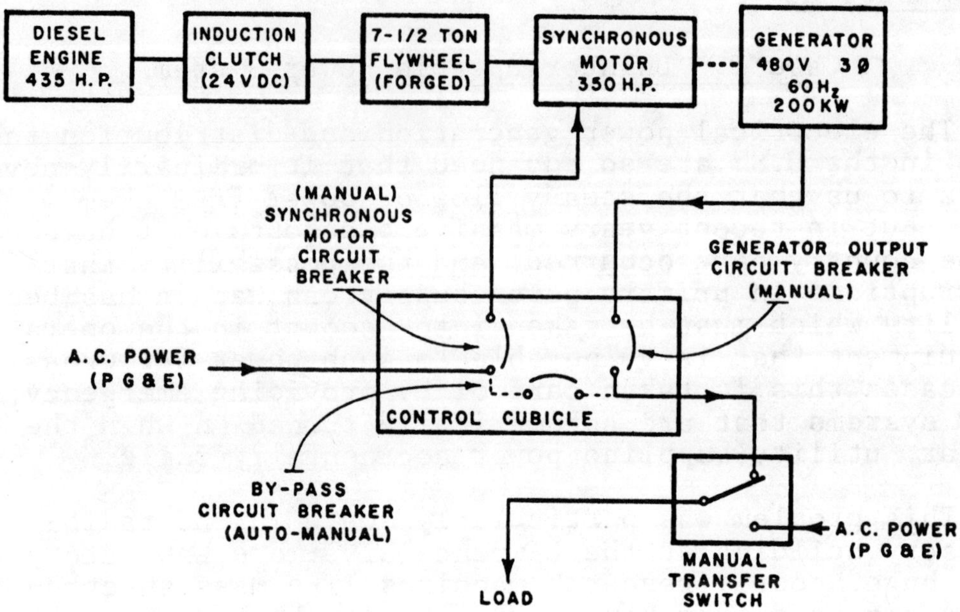

Fig. 3-1. Functional Schematic, LLL 200-kW No-Break Power System.

then drives the motor-generator shaft and thus supplies the needed emergency power (Fig. 3-1).

To meet LLL's rigid requirements for constant frequency and voltage during the change-over to emergency power would require a flywheel of very high energy-storage capacity. This high energy-storage capacity could be provided by a flywheel of huge size running at a low conventional speed. Or it could be provided by a flywheel of moderate size running at a much higher speed. However, cast steel normally used for manufacture of flywheels would burst at this high speed, so the LLL engineers specified the smaller flywheel, running at the high speed, but specified that it be made of forged, not cast, steel. The forged steel is strong enough to withstand these high rotational speeds.

The vendor who got the bid to build this system thought that the conventional cast steel flywheel would be strong enough so furnished the motor-generator system with such a flywheel. The engineers at LLL rejected it and it had to be transported back across the country and replaced with a forged steel flywheel. This system has been in operation successfully for years. It thus furnishes a new product to be manufactured and sold by producers of emergency power systems in the high output range of around 200 kw (1)(2).

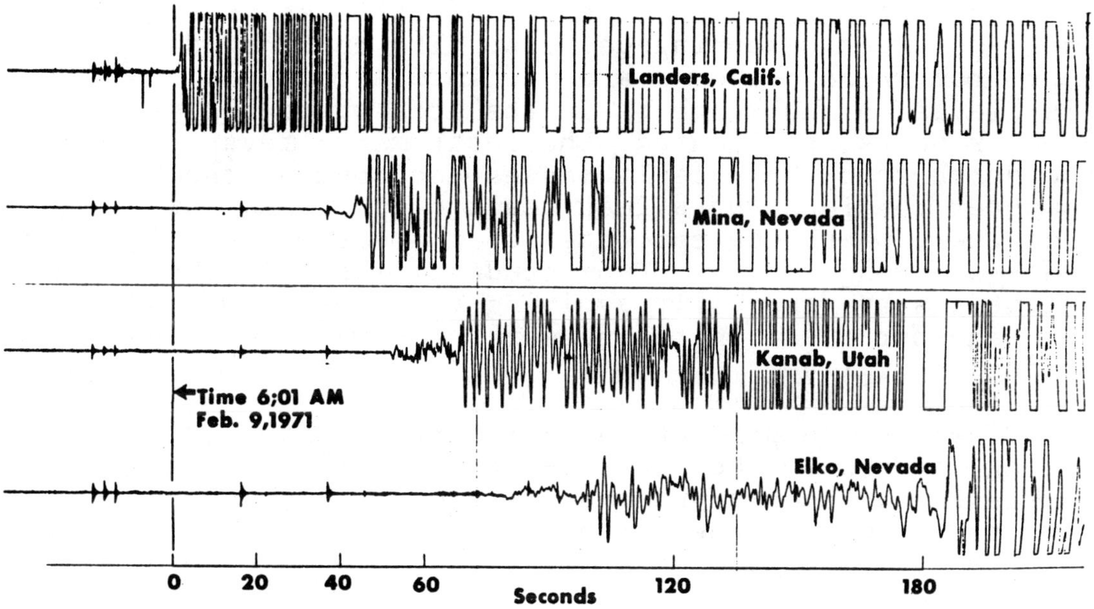

Fig. 3-2. Record of destructive L.A. earthquake as recorded at Seismic Data Center at instant it happened (6:01 a.m., Feb. 9, 1971). Time difference in arrival of shock (full scale wave amplitude) at different sensing stations permits calculating location of quake center.

The engineer at LLL who developed this system was so occupied with other developments he had to make in the radioactive chemistry building that he never got around to writing up this system for publication. To save it from being lost to industry, the technical communicator assigned to work with this engineer produced a write-up on it which was published in a technical magazine. That article is reproduced in its entirety in Appendix 3-1.

Case II. Remotely Controlled Seismic Network

Earthquakes used to be looked on as entirely an "act of God", hence unforeseeable and unpredictable. But in the last few decades earth scientists have been moving closer to the idea that the occurrence of an earthquake is the result of a long chain of transformations in the earth's structure which obey certain laws of earth science and therefore their occurrence might be foreseen in advance. However, such advance knowledge of when an earthquake will occur and where depends on having a great mass of data on very slight earth motions that occur in earthquake-prone areas over a long period of time. Scientists have therefore developed a whole technology of earth-shock and earth-movement detection and measurement instrumentation. Most of this instrumentation consists of networks of isolated earth-movement measuring devices. The combined data outputs from these networks are fed into a computer. When a sizable earth-shock occurs in the area where the seismic network units are located, the instruments measure and time the earth-shock wave as it passes each one of them. The computer combines these instrument outputs and computes, from them, where the epicenter of the earth-shock is located and how strong it was. From studies like this the earth scientists are gradually evolving the know-how to predict when another quake in the network area will occur and how strong it will be.

The precise prediction of future earthquakes is still a long way off in the future, but meanwhile government bodies in earthquake-prone areas are anxious to get all the data they can from the slight tremors and earth movements that are constantly occurring, so seismic instrumentation networks are in demand and already constitute a valuable line of commercial products.

When the testing of nuclear weapons was taken out of the atmosphere and moved to underground test shots, LLL found a need for seismic networks to study these shots. They evolved such a network that has valuable technological characteristics. These are:

1. The seismic instruments are all automatic and completely remotely controlled from one central point. It is not necessary to send out operators to calibrate, adjust, and test the instruments. This is all done by an operator sitting at one central point--in this case the LLL at Livermore.

2. The seismic instruments are all of the same standard design, so standard units can be built in quantity, thus lowering the cost.

3. The seismic instruments are all connected to the central control point by commercially available telephone lines, which can be leased in almost any part of the country. Seismic networks can then be set up in any area of the country quickly, and at low cost, because they can all be connected by telephone lines available everywhere.

One such network was built by LLL to measure the earth-shocks generated by the nuclear explosives it sets off underground. These measurements give the LLL a good indication of the explosive ability of the nuclear device that was set off in the underground tests. This network therefore gives LLL a quick and easy means for measuring the explosive power of a nuclear bomb--the most important characteristic sought by the bomb designers.

However, this system will pick up any other earth-shocks that may occur in the seismic network area, so it can provide very useful data on earthquakes occurring in that area. This proved to be valuable when a sizable natural earthquake occurred in the nearby Los Angeles area in 1971. The records of this earthquake produced by the instruments of the LLL seismic network, shown in Fig. 2 of Appendix 3-2, gave prompt and accurate data, from which its epicenter was quickly deduced.

This network has also demonstrated another fringe benefit. When the USSR set off a "monster" atomic bomb in an underground test in Siberia, the LLL seismic network picked up the resulting shock wave, as shown in Fig. 3 of Appendix 3-2. These records enabled the scientists at LLL--the U.S. nuclear weapons design laboratory--to determine the size of the Soviets' nuclear bomb and where it was set off.

Now that Government bodies are interested in buying seismic networks with which they can monitor the earthquake hazards in their respective areas, this technological development by LLL affords a potentially valuable commercial product that can be achieved by transferring it to commercial application. However, no write-up at all was produced on it that would enable industry to know of it and understand it. The technical writer/editor assigned to the group of engineers that produced this development felt it should not be lost to industry, so produced and published the write-up presented in Appendix 3-2.

Case III. Transportable Nuclear Explosives-Test Instrumentation System

For years LLL carried on a huge R&D program for advancing the use of nuclear explosives in civilian, not military, projects, such as the building of dams, canals, harbors, etc. One particularly promising project under this program was the use of underground explosions of atomic bombs to loosen up oil-containing geological formations and thus increasing the output of oil from wells in that area. This required an elaborate instrumentation complex that could detect and measure the effects of the underground nuclear explosion comparable to the instrumentation system already installed at the military nuclear explosives test site in Nevada. The instrumentation system at this military nuclear explosives test site is a permanent installation, but for the oil-output-increasing experiments it was necessary to have an instrumentation system that was transportable from one promising geological formation to another, often many hundreds of miles away. LLL therefore developed a whole technology of transportable instrumentation for exploring the effects of underground atomic bomb explosions similar to the fixed instrumentation installation at its test site where atom bomb tests for military purposes were conducted.

The extensive and diverse technology evolved for this transportable instrumentation system comprised many individual technological developments of value to industry, such as microwave communication links, remotely controlled power supplies, supervisory remote control systems, etc. As occasionally happens in the case of research facilities construction, the typical contractual R&D reports were not produced. This left a large complex of valuable technology lost to industry because it was not written up and publicized so that industry would know about and understand it. However, a brief summary of this entire transportable instrumentation technology was prepared as an internal report (3) to give the EE Department, which developed it, a record of what had been accomplished. Using this summary as a base, a technical writer/editor associated with the EE Department prepared and published the write-up on this technology which is reprinted in Appendix 3-3.

Summary

The Government R&D programs are funded primarily to achieve desired research results. This funding calls for R&D reports that describe the results achieved, which we call "contractual reports." These contractual reports are rigidly required by the funding agencies, so are always produced. Such results are often published, so the technology they contain becomes widely known to and understood by industry, hence is frequently transferred to commercial use.

But a large part of the R&D program funding is devoted to developing the facilities required to achieve the desired results. For the technology developed in these facilities, R&D reports explaining what was accomplished and how it was achieved are often not a primary requirement, so may not be produced. Much of this technology therefore is not written up and published, or it gets written up only in meager internal reports. A surprisingly large amount of the technological development contained in the R&D facilities built to execute R&D programs therefore does not become known to and understood by industry, and therefore is not available to industry through the technology transfer process. The technical writer/editors of each R&D center can therefore make an enormous contribution to the increase of technology transfer by digging up these neglected technological developments and producing write-ups on them which then get published.

REFERENCES

(1) The Uninterruptible Power Evolution: Are Our Problems Solved? Robert L. Knight and Moon H. Yuen, IEEE TRANSACTIONS ON INDUSTRY APPLICATIONS, Vol. IA-9, No. 2, March/April 1973.

(2) Uninterruptible Power Supplies in the Textile Industry. L. J. Boutin, IEEE TRANSACTIONS ON INDUSTRY APPLICATIONS, Vol. IA-9, No. 2, March/April 1973.

(3) Rio Blanco Electronics, O. H. Krause, J. B. Carson, M. F. Jaroska, S. J. Spataro, and G. E. Russell. UCRL Report 51501, April 8, 1974. Available from NTIS, Springfield, VA 22161.

Ch. IV

GIVE A HISTORICAL PERSPECTIVE ON THE WHOLE TECHNOLOGY

Outline

	Page
Specialized Technologies	29
Case I - Low Level Radiation Measurement	30
Case II - Time Microscope - The Ultra High Speed Cathode Ray Oscilloscope	33
Case III - Time Domain Reflectometry - Cable Length Measurement	36
Time Domain Reflectometry Principle	38

Appendices

App. 4-1	Low Level Radiation Measurement	112
App. 4-2	Precision Cable Length Measurement	116

Ch. IV

GIVE A HISTORICAL PERSPECTIVE ON THE WHOLE TECHNOLOGY

Specialized Technologies

Government R&D programs rarely create a single, isolated, unique technological development. Rather, they produce a whole cluster of interrelated technological developments which, in combination, produce a whole new specialized (although limited) technology.

For example, in the R&D program conducted at LLL to develop high power lasers, the laser beam is produced by first generating a low power beam of the desired wave form and frequency in a precision oscillator. Then this low power beam is built up to very high energy by passing it through a chain of laser amplifiers. The minute amounts of misalignment, as the beam passes from one amplifier stage to the next, cause some of the powerful beam to be reflected backward. This backward-reflected energy can damage and even destroy the glass optical elements of the laser system. So a whole group of technological developments were produced which, in combination, eliminate this damage--beam polarizers, beam rotators, beam cross section controllers, damage resistant glass for the optical elements, etc. Thus a whole specialized technology of laser self-damage-elimination was produced in this R&D program.*

When industry looks through Government R&D programs for valuable technology to transfer to commercial application, it is especially interested in these new, specialized technologies because each one can lead to a specialized area of industry in which the company can become proficient and thus make it profitable.

To arouse industry interest in these new, specialized technologies one must prepare and publish an article on each one that gives a historical perspective on the whole new technology. That is, the article must tell how one of the component technological developments of the new, specialized technology leads to the next until the whole technology has evolved.

---

*Many laboratories are engaged in a great variety of laser developments, so are all more or less engaged in solving this laser self-destruction problem. But the enormous power of the LLL lasers makes this problem most acute there, so LLL has concentrated heavily on its solution and played a lead role in creating the technology for it.

These historical perspectives can be of different types, based on the different development patterns by which the new technology was created. For example, the new technology may comprise a number of functionally related technological developments, each of which leads to the next one until the whole technology is evolved. This perspective is presented in Case I. In other cases the historical perspective shows how progressive stages of advance are produced in a single line of equipment by applying a new, more advanced, physical principle to accomplish each stage of advance. This type of historical perspective is demonstrated in Case II. Still another type of historical perspective is based on the fact that a succession of progressive stages of automation is produced in a single type of equipment, resulting in a whole specialized technology for such equipment. This pattern is demonstrated in Case III. In short, the different types of historical perspectives are demonstrated by the case histories which follow.

Case I. Low Level Radiation Measurement

During the years when atomic weapons were being tested in the atmosphere (1947 to 1963) the technology of high level radiation measurement became highly developed. These test shots produced radioactive debris which consisted of highly radioactive isotope materials. Each isotope produced a characteristic spectrum of different radiations at different intensities. When these isotopes were mixed in different proportions in the debris, the resulting emission of nuclear radiations was a complex mixture of different alpha rays, beta rays and gamma rays, of widely varying intensities.

By the time atmospheric tests were brought to a halt the technology for unravelling and measuring these complex radiation emissions had been pretty well developed. It consisted of two major technologies--radiation detectors and pulse height analyzers.

Radiation detector technology had achieved devices which generated an electrical output when they were subjected to incident radiation. These outputs were in the form of electrical voltage pulses. But these outputs were not continuous. Instead, the incident radiation produced a series of rapid voltage pulses. The rate at which these pulses were produced corresponded to the intensity of the incoming radiation. Also, the magnitude of each pulse corresponded to the frequency or energy per photon ($E=h\nu$) of the incoming radiation.

The pulse height analyzer was an electronic instrument which sorted the pulses according to their voltage level, and totaled the number received of each level. Since the voltage levels of the pulses correspond to the photon energies or frequencies of the incident radiations, the pulse height analyzer drew a curve, on a cathode ray tube screen, showing the photon energies or frequencies of the incident radiations on the horizontal axis and the intensity of each frequency on the vertical axis. Since each isotope in the radioactive debris had a sharply characteristic emission frequency, this showed up as a sharp peak on the intensity-frequency curve on the cathode ray screen (Fig.4-1). Such pulse height analyzers had started out originally with such poorly defined peaks as occur on the dotted line in Fig. 1 of Appendix 4-1, but had been developed to the point where each isotope in the radioactive debris was identified by a sharp peak, as shown by the solid curve in that figure.*

About the time the atmospheric tests ended the people making the bombs became concerned with what happened to the world as a result of the fallout from bomb tests. The fallout was of very low, rather than of very high, radiation intensity, so they needed an entirely new low-level radiation measurement technology to find out how these fallouts affected people, animals, vegetation, etc.

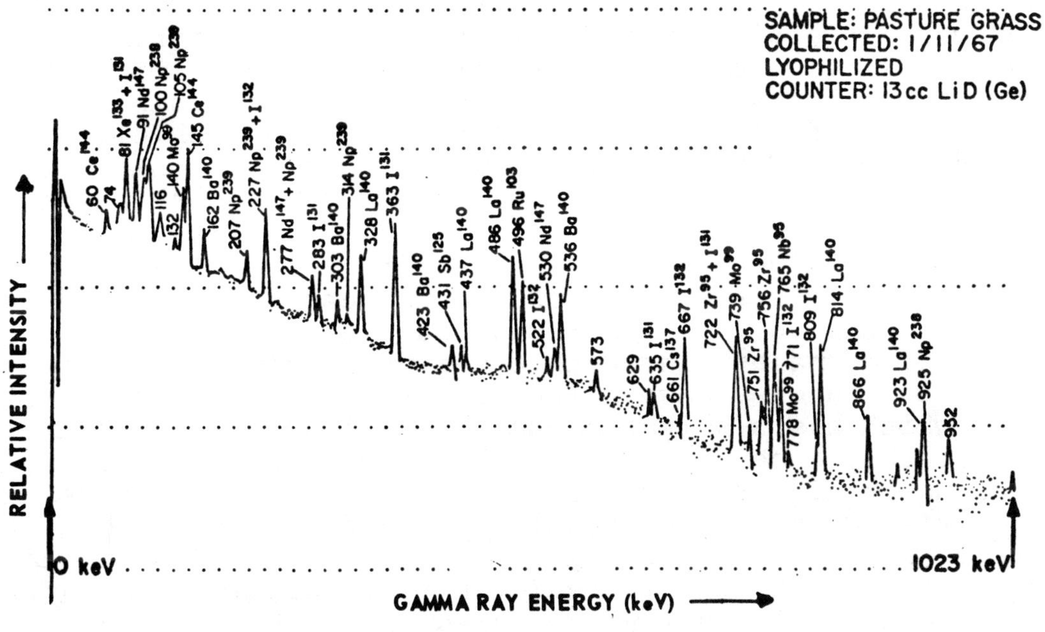

Fig. 4-1. Local pasture grass (in Livermore, CA) contaminated by fallout from Chinese bomb explosion. Spectrum was produced from input to a 13 cc planar Ge(Li) detector.

*For details see Appendix 4-1.

The Lawrence Livermore Laboratory (LLL) therefore set up a biomedical department and, to enable this department to study the effects of fallout, a program to develop low-level radiation measurement technology was started.

The first problem was to develop radiation detectors that were sensitive enough to develop usable electrical pulse outputs even from extremely low intensity radiation inputs. A then newly developed solid state detector, comprising a layer of lithium diffused into the surface of a germanium crystal, afforded the sensitivity required. But reliable detectors of this type could not be manufactured because in some germanium materials the lithium did form a diffusion layer, in some it did not. So a development program was undertaken which solved the problem and made consistently good lithium-drifted germanium detectors possible. LLL then became a center to which other laboratories flocked to learn this technique.

This led to another problem. The levels of radiation from fallout were so low that they were masked by the prevailing background radiations present everywhere as result of minute amounts of radioactive isotopes present in the soil and from other natural sources. The problem then was to devise shielding in which to enclose the entire low-level radiation measuring apparatus so as to eliminate these stray background inputs. To accomplish this they first enclosed the apparatus in a thick steel shield similar to an old-fashioned office safe. But this only intensified the background radiation instead of diminishing it.

An extensive program of scientific detective work followed to uncover the cause of the increased background from the steel shield. It was found that in modern steel mills the blast furnaces, in which the iron is first extracted from the iron ore, have a furnace wall that is made of fire brick which has a slight amount of radioactive material deliberately embedded in it. This produces radioactive emission outside the furnace wall which can be measured. When this emission output falls off, it indicates that the firebrick lining is getting too thin and should be replaced. But in the meantime this radioactive material in the brick contaminates all the iron produced in the blast furnace. So the solution was, make the shield of steel not made in modern blast furnaces. Fortunately, steel salvaged from the ships sunk at Pearl Harbor provided the answer and a shield produced of that material did effectively reduce the background radiation present around the low level radiation measuring apparatus.

Thus the historical perspective here is that one technology development led to another functionally related one until the entire new specialized technology was evolved. A comprehensive article giving details of this new technology was prepared by the engineers and physicists involved in it and published in a scientific journal (1). But it was so technical that only the specialists in the field could read and understand it. To arouse widespread industrial interest in this new, specialized technology, the writer/editor assigned to work with the electronics engineers who made the development wrote an article which gave the full historical perspective on it. This article was published and is reproduced in its entirety in Appendix 4-1.

Case 2. Time Microscope - the Ultra-High Speed Cathode Ray Tube Oscilloscope

Another common pattern of technological development in the evolution of specialized new technologies is that in which a single type of equipment moves to progressively higher stages of technological advance. For example, in the case of giant main-frame digital computers, the first type were the Univacs, based on the technology of vacuum tubes. The next stage in their technological advance was ushered in by the LARC (Livermore Advanced Research Computer, built by Sperry-Rand for the Lawrence Livermore Laboratory). It was based on the technology of transistors (solid state devices) that superseded vacuum tubes. Since then we have giant mainframe computers based on the more advanced technology of integrated circuits, and looming on the horizon are mainframe computers based on the still more advanced technology of fiber optics.

However, these are all progressive stages of technological advance that were achieved in and by different R&D organizations. Of greater interest in the promotion of technology transfer are cases of this pattern of technological development in the evolution of new, specialized technologies in which the successive stages of technological advance in a single line of equipment occur at the same R&D center. Such a case is afforded by the technology of ultra-high speed cathode ray tube oscilloscopes, the successive stages of which occurred in a single R&D center--LLL.*

A basic need in scientific and engineering research and development work is to produce a record of what takes place. How did pressures, temperatures, velocities, etc., especially of moving objects, change during the test or experiment?

---

*Although this whole technology was developed at LLL, it incorporated advances then being made in electronics firms (RCA, DuMont, Tektronix), and in other major laboratories.

The universal solution to this problem is to use a "transducer" which converts the variable to be measured (pressure transducer, temperature transducer, acceleration transducer) into a corresponding electrical output. Until World War II it was universal practice to feed these outputs into a magnetic oscillograph--an electrical meter which used a light beam as a pointer. With such a lightweight pointer the meter element could move very rapidly to respond to fast changes in the electrical output being measured. The light beam pointer impinged on a photographic chart paper and thus gave a recorded time curve of the variable being measured--whether pressure, temperature, etc.

This was fine for phenomena that occurred in milliseconds or longer time intervals. For any phenomena taking place in time intervals faster than a millisecond, the magnetic oscillograph was too slow. During World War II the development of radar made it necessary to measure the time it took for a radio pulse to travel to the target and return--at 186,000 miles per second. By then the cathode ray tube oscillograph had been developed. This used a beam of electrons as a light beam pointer. The beam was moved by voltages applied to deflection plates through which the electron beam passed. This was a more sensitive meter of electrical outputs than the magnetic oscillograph, and could record phenomena that occurred in as little as a micro-second--a thousand times faster than the magnetic oscillograph--fast enough to show the time it took the radar pulse to go to the target and back by the spacing between the peaks on the face of the cathode ray tube.

In atomic weapons work many phenomena occur in microseconds or less, so the cathode ray tube was adopted as the principal device for recording these phenomena by converting them into electrical outputs (through transducers) and recording these outputs by photographing the curves they produced on the faces of the cathode ray tubes. But the atomic weapons tests imposed requirements that were too severe for the cathode ray tubes used in radars to meet, so the electronics engineering department of the Lawrence Livermore Laboratory started a chain of advances in cathode ray tube oscilloscopes to meet its requirements.*

An article giving a "historical perspective" on these developments (see Ref. (2)) points out the major problems as follows.

"In the first place, the signal to be measured is, in the great majority of cases, a one-shot pulse with the pulse occurring in a matter of microseconds,

---

*See footnote, p.33.

often in only nanoseconds or less (thousandths of a millionth of a second). This requires cathode ray tubes and the appropriate timing and triggering and (sometimes) amplifying circuits that go with them which are also capable of extraordinarily fast response. The entire combination of tube and related circuits is called an "oscilloscope system." To indicate how fast a response the oscilloscope system will have to have, the engineers speak of the range of sine wave frequencies it will have to be able to pass - commonly called its "bandwidth."

"Another complication is that the oscilloscope system must measure electrical voltages over a tremendous range of magnitude - from millivolts to thousands of volts - what engineers call a huge "dynamic range." In addition, the tube must have an electron beam intense enough to provide fast writing speed. Furthermore, it must permit superimposing a sine wave timing signal on the experimental data curve.

"Besides these extraordinary technical requirements, the oscilloscope system must meet some very difficult practical ones. Since these oscilloscopes have to be set up in bunkers or trailers, it is very important that they be small in size and that they consume very little power. And above all, they must be reliable. Weapons tests are extremely expensive, and it would be costly indeed if an oscilloscope failed to function as intended during a shot and the desired data were lost."

The chain of technological development of cathode ray tube oscilloscopes began at LLL in 1954 with the L-1 design. An improved design, the L-2, was produced in 1961. As noted in Ref. (2);

"That brought the LRL advances in electron beam oscilloscope technology to the limit of what could be done directly, by deflection of an electron beam with a voltage applied to fixed deflection plates, so the trend of development shifted to traveling-wave type oscilloscopes which made use of the traveling-wave (TW) cathode ray (CR) tube already developed by J. R. Pierce of Bell Labs and others. In these advantage is taken of the traveling-wave principle. That is, the beam is deflected, not by a single set of deflection plates, but by a series of sets of deflection plates, and the deflection voltage moves progressively along these (in a traveling wave) at the speed at which the electrons in the beam are traveling. This provides much more powerful deflecting action with a given deflection voltage. In effect it produces a much faster mechanism for moving an electron beam across the tube face, hence a substantial increase in bandwidth.

"When development of the traveling wave tube reached the limit of speed in oscilloscopes that could be obtained with that principle, a new technical advance opened the door to a new plateau in oscilloscope operating characteristics. This was the development of "fiber optics." A fine glass thread or fiber produces total internal reflection. That is, light which enters one end of the glass fiber will be conducted, like water in a pipe, through the length of the fiber until it comes out at the other end. This occurs regardless of how long the fiber or how curved it is. A bundle of such fine fibers thus acts as a "light duct" to conduct an image, say the one on the face of a cathode ray tube, the way a waveguide conducts an electromagnetic wave.

"This principle is applied to push the characteristics of a cathode ray tube to still faster performance in this way: In place of a lens system to transfer the image on the cathode ray tube (CRT) face onto the recording film in a camera, a fiber optic plate is mounted on the tube face to do that. The lens system has a light transmission efficiency of less than 10%, but the fiber optic face plate has a transmission efficiency over 60%. Hence much more of the light generated by the CRT beam is used to act on the film, in effect producing a substantial increase in writing speed."

A cathode ray tube oscilloscope based on this principle was also designed and was in practical use for years (the L-11 design). It had considerably better characteristics than the tubes based on the traveling wave principle. As concluded in Ref. (2):

"One can thus see how the development of faster and faster oscilloscopes for nuclear test instrumentation has proceeded from one plateau to another, each successive plateau made possible by the application of some new technological principle as fast as it appeared on the horizon." (See Fig. 4-2 and Ref. (3)).

Case 3. Time Domain Reflectometry - Cable Length Measurement

In a great deal of modern research and development work, especially in the atomic energy field, the phenomena

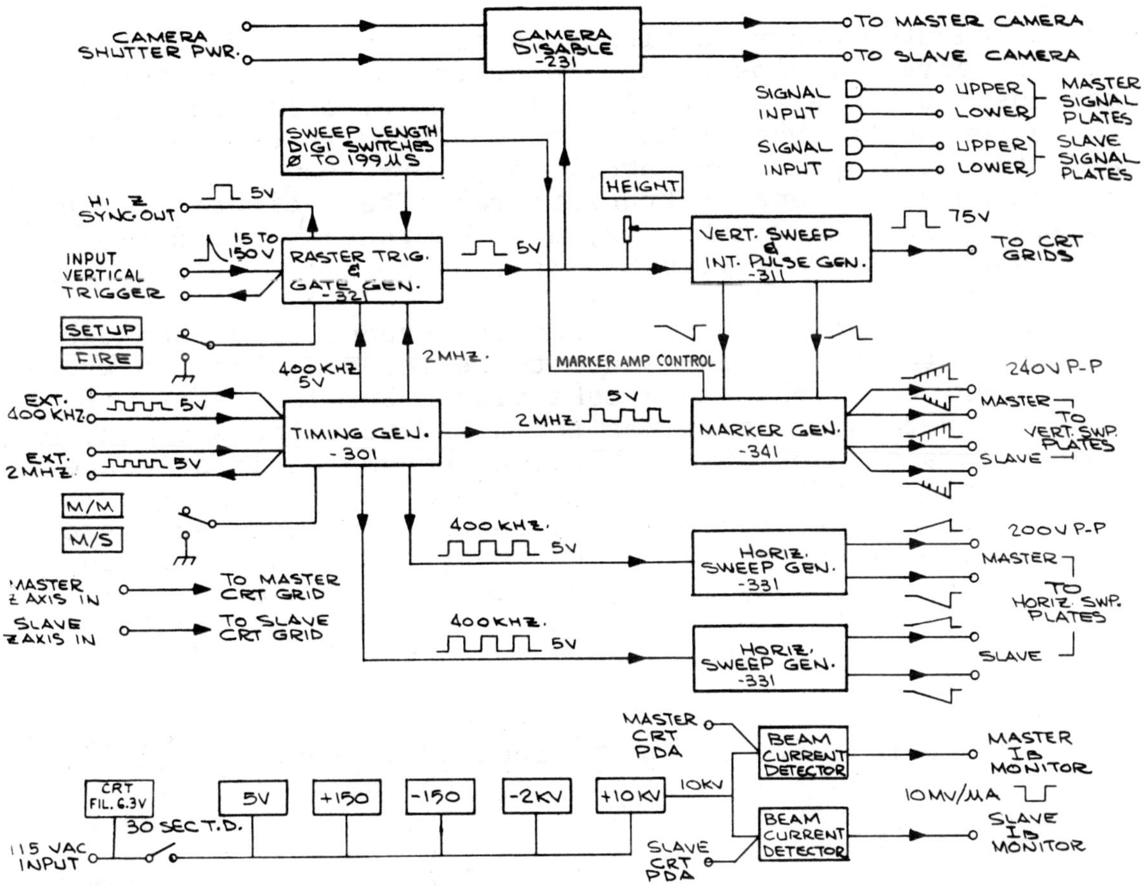

Block diagram of L-300 Raster Oscilloscope

Fig. 4-2. Included in the LLL-developed high-speed cathode-ray oscilloscope technology is a series of successive designs of a raster-type oscilloscope system for precision measurement of microsecond time intervals. The series began with a design using vacuum tubes. It included a few racks of equipment. This was followed by a design (the L-10 Raster Oscilloscope) which used transistors instead of vacuum tubes (3). It was only desk-top size and had better operating characteristics. Lately a new design (L-300) was produced (1977) which uses the more advanced technology of digital instead of analog time measurement systems and uses integrated circuits instead of individual transistors. This makes a smaller unit with better performance characteristics. (See Ref. (4)).

to be measured occur in microseconds or even much shorter time intervals (nanoseconds -10^{-9} sec., or even picoseconds -10^{-12} sec.). It is universal practice to convert the experimental variables to be measured (pressure, temerature, light emission, etc.) into corresponding electrical outputs by means of transducers. These electrical outputs, as explained in Case 2, are recorded by the use of high speed cathode ray tube oscilloscopes.

However, in the case of atomic weapons tests and other research, the transducers have to be located close to where the experimental phenomena take place, but for safety reasons the recording apparatus to which the outputs of these transducers are fed must be located considerable distances --even up to hundreds of yards--away. This produces a delay in the time of arrival at the recording apparatus of the electrical outputs generated by the transducers. In view of the extremely short duration of these electrical output pulses, this arrival time delay can cause serious errors in the data recorded by the cathode ray oscillographs (5).

This delay time is much reduced by sending the electrical outputs over coaxial cables instead of over ordinary 2-wire circuits. Therefore dozens, even hundreds, of coaxial cables are used in many experimental R&D setups.* To keep the experimental results accurate, the exact time it took an electrical output pulse to travel from the transducer, where it was generated, to the cathode ray oscillograph, where it is recorded, must be known. Since electrical pulses travel over coaxial cables at a practically constant velocity, knowing the length of coaxial cable over which the electrical output pulse traveled makes it possible to compute exactly the time it took the pulse to go from the transducer to the recording unit. However, the electrical pulses travel at speeds around 186,000 miles per second, so to get appreciable accuracy in travel times means having to measure coaxial cable lengths to extreme accuracy. This cannot be done mechanically, but can be done electrically by a technique called "time domain reflectometry."

Time Domain Reflectometry Principle

This technique began in the days when telegraph cables were first laid across the Atlantic Ocean. If the cable broke, it was important to find where the break occurred, without pulling up the cable to examine it. The people who

---

*Coaxial cables are also used because they provide shielding and reduced attenuation of high frequencies.

laid the cables knew that an electrical pulse, applied to the cable at the land end, would be reflected back when it hit an open end of the cable--that is, where there was a break. The reflected pulse could be easily detected when it returned to the input (land) end. If the time it took the electrical pulse to travel down the cable to the break (open end) and be reflected back could be accurately measured, the distance to the break could be easily calculated.

This technique is used in time domain reflectometry. An electrical pulse is applied to one end of the coaxial cable length to be measured. The time when this pulse leaves the input end is marked by a peak on a time measurement cathode ray tube such as is used in radars. The time when the reflected pulse comes back to the input end is indicated by another peak on the cathode ray tube screen a short distance away. The distance between the peaks measures the time interval between when the pulse started down the cable and the time it returned. This is just twice the cable length. From this time measurement the cable length can be easily calculated.

However, this principle of cable length measurement poses problems. The time intervals to be measured are in microseconds. Furthermore, as noted in Appendix 4-2.

> "The sensitivity of the time measurements has to permit clear indication of 1/4 of an inch in 10,000 feet of cable, or a resolution of one part in 500,000!"

To cope with these problems the EE Department at Lawrence Livermore Laboratory developed progressively better pulse travel time measurement technology, based on increasingly automated apparatus for performing the same basic time measurement technique. This whole specialized technology was described in an article which gives a historical perspective on its evolution and thus makes the whole technology understandable to industry. That article is reproduced in its entirety in Appendix 4-2.

As indicated in that article, the progressively more automated facilities for performing the same time measurement technique comprised these successive stages of automation:

> "To cope with these problems, progressively better pulse travel time measurement techniques were developed. The first used the standard laboratory technique of inserting calibrated delay elements, in place of the cable to be measured, in the measuring-circuit until signals were matched on an oscilloscope screen.

"The concept which paved the way to really successful pulse travel time measurement was the idea that, instead of using the start and reflected pulses to show up as peaks on an oscilloscope, they be used to start and stop a flow of high frequency (100 megacycle) pulses generated by a local oscillator. The number of pulses passing through between the start and stop signals could be counted on the newly available high-speed electronic counters, thus giving a very precise, digital measurement of the time it took the pulse to travel to the end of the cable and be reflected back.

"This sequence of operations was made automatic by a one-tube circuit arrangement devised by Ed Laine (6) which Bob Werner further developed so that it gave a direct, visual reading of the desired cable length (one way pulse travel time over the cable being measured) in nanoseconds. Werner later developed a small transistorized version of the same unit. This made available a "cable calibrator" device which gave immediate, automatic cable length measurement readings, to the accuracy (resolution) desired, and which was available in a small handy package. UCRL 7078, mentioned above, gives full details of this development. A self-contained field-use instrument, with its own power supply and counter, was developed by Tom Holdsworth and John Glasgow (LE 13807-1).

"To bring the cable length (pulse travel time) measurements up to the range and resolution required, it is often necessary to use a "bias" cable in series with the cable to be measured. After the measurement of the unknown-plus-bias-cable is made, the known length of the bias cable has to be subtracted from it, to give the actual length of the cable being measured. For cable measurements made in the field this was a nuisance, so the "cable calibrator" was improved so that it would do this subtraction automatically, by adding to it a miniature computer and a memory.

"With this unit ("cable calibrator, auto subtract," LE 15313) any technician, with a few minutes training, can instantly make the measurements which formerly required an engineer and considerable time to do. Also this unit costs $3,000 compared to the $11,000 for the initial cable length measuring equipment. So successful has this system been that it is now in wide use by most AEC contractors (Sandia, EG&G).

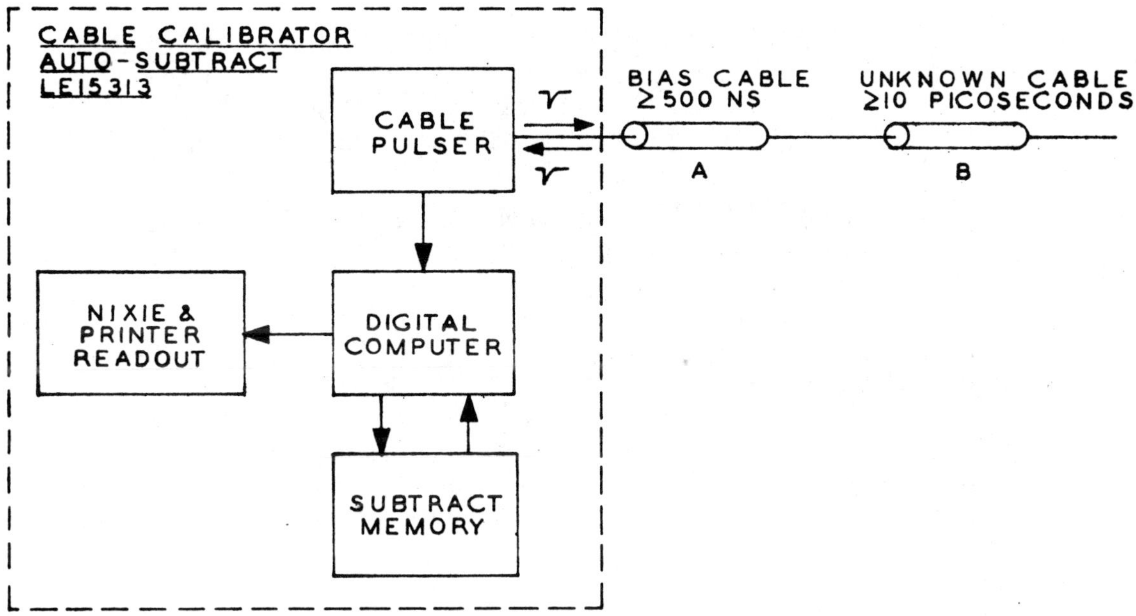

Fig. 4-3. Functional block diagram for automatic cable length measurement, Cable Calibrator Auto-Subtract.

To conclude: Industry is very much interested in the new specialized technologies created in Government R&D programs. The technical communicator can stimulate this interest by writing "historical perspective" articles on those specialized technologies created in the R&D center where he is employed. The techniques for doing so are presented above.

REFERENCES

(1) Gamma-Ray Spectrometers for the Assay of Complex Mixtures of Low Concentration of Radionuclides in Environmental and Biological Materials. Paul L. Phelps. UCRL Report 70544, Oct. 24, 1967. Available from NTIS. Also published in IEEE TRANSACTIONS ON NUCLEAR SCIENCE, Feb. 1968, Vol. NS-15 No. 1, pp. 376-382.

(2) Time Microscope - The Ultra High Speed Cathode Ray Oscilloscope. ELECTRONICS ENGINEERING DATA LETTER, Vol. XII, No. 10, Oct. 1969, pp. 1-4. Issued by Electronics Engineering Department, Lawrence Livermore Laboratory, P. O. Box 808, Livermore, CA 94550.

(3) Use RASTER OSCILLOSCOPES for Faster Time Measurements. R. P. Rufer and Wm. A. Karlotski. ELECTRONICS, Dec. 28, 1962. Copyright 1962, McGraw-Hill Publishing Co. Inc.

(4) L-300 RASTER OSCILLOSCOPE OPERATION AND MAINTENANCE MANUAL. J. C. Lawson, M-080, Lawrence Livermore Laboratory, Feb. 2, 1977, 116 pp. Fig. 4-2 is reproduced from Fig. 10, p. 27, of this Manual.

(5) FIDUCIAL DISTRIBUTION SYSTEM, Laurence P. Altbaum. Lawrence Livermore Laboratory Report UCID 15336, June 1968, 17 pp.

(6) Getting Subnanosecond Precision in Coax Cable Delay Measurements, E. F. Laine. ELECTRONICS, Feb. 1, 1963. Copyright 1963 by McGraw-Hill Publishing Co. Inc.

PART II

FINDING THE TECHNOLOGY TRANSFER INFORMATION

Ch. V

PROVIDE A GUIDE THROUGH THE MAZE IN YOUR OWN GOVERNMENT AGENCY

Outline

	Page
The Maze	43
The Guide	44
Content	45

Ch. V

PROVIDE A GUIDE THROUGH THE MAZE IN YOUR OWN GOVERNMENT AGENCY

The Maze

To pick up a technological development that is worth transferring, from a Government R&D program, is easy enough for a person in industry to do. You simply read about it in a technical magazine or conference proceeding. But to convert that technological development into a commercial product is a difficult engineering job. To do it economically you have to get, from the Government R&D center that created it, all the technical know-how they acquired in the process of producing that technology development. You cannot afford to go reinventing the wheel.

For example, after an article on the above-mentioned electron beam welder (Ch. II, Ref.7) developed at LLL was published in an electronics magazine, a number of firms and individuals wrote to that agency for further information about it. Several wanted to know how to get sets of detailed drawings on it; one wanted to know whom he could talk to about building a very long, narrow, high-vacuum work chamber needed to weld fuel rods for a nuclear reactor. Still another pointed out that the welding process causes the emission of metal vapor into the evacuated work chamber and this could destroy the high electron-emission characteristic of the cathode which generates the electron beam. Whom could he talk to about overcoming this problem?

The result is an enormous demand, by the people in industry who want to do technological transfer, for the technical information needed to effect such transfer. Naturally they seek this technical information from the R&D centers in which the technological developments they want to transfer were created. As a result most of the Government R&D centers or agencies have set up "offices of technology utilization" or "technology transfer centers" to satisfy that demand. For minor technological developments this is adequate but for effective transfer of major technological developments, such as the electron beam welding machine discussed in Ch. II, this is not enough. To transfer substantial Government-produced technological developments to meet the commercial needs of an industrial firm it is vital to have a direct, personal contact between the person seeking to acquire a Government-produced technological development and the person in the R&D center who created it.

Furthermore, it is not enough to develop such a direct contact with just a single person at that R&D center because, in most cases, the technological development was produced as a result of different steps in the R&D process being performed by different groups in the R&D center.

For example, consider a new instrument developed at LLL for making surveys, in the field, of radioactive contaminants that may be present. This instrument (1)(2), makes possible sensitive, precision measurements not possible with field survey instruments before, so it would be a valuable product for some firm to produce commercially.

It was developed as a project of the Hazards Control Department at LLL, so to effect its transfer to commercial application you would have to have direct contact with the people in that Department who were engaged in its development. But once you contacted them you would find out that they only developed the radiation detectors used in this instrument and the case for making it portable. The pulse height analyzer, which makes its precision measurements possible, and the power supply for the whole instrument--in other words, all the electronics--were conceived and developed by the Electronics Engineering Department. So, to effect transfer of this device into a commercial product you would have to develop direct personal contacts with the people in that Department also.

Thus, to convert a promising Government-produced technological development into a successful commercial product, the outsider must establish direct personal contact with all those in the R&D center who were involved in its creation. However, to establish such a connection with the personnel inside a Government R&D center presents the outsider with this immense obstacle. Government R&D centers are organized for doing research and development, not for doing technology transfer. Therefore, to the outsider trying to find his way around in it, the R&D center is a huge, bewildering maze because of its huge size (thousands of employees); the vast complex of interrelated technical activities going on in it; and its usual practice of storing all the technical information generated in that R&D center in hiding places to which the outsider is not allowed access.

The Guide

It is therefore vital that, to help outsiders to benefit from the technological developments it creates, each R&D center should produce a guide, through its own maze, that will help the outsider to achieve, quickly and easily, the direct personal contacts he needs to effect the technology transfer he wants to achieve for his firm.

The most valuable contribution that the writer/editor employed in a Government R&D center can make to the increase of technology transfer is therefore to produce such a guide, for the R&D center he is employed in, if it does not already have one. A technique for producing such a guide can be readily evolved by keeping in mind certain considerations about its content that are presented below.

Content. To provide a guide through the maze of your own organization, that the outsider seeking technology transfer information can follow, requires giving a clear picture in each of these three areas.

1. Organization.

2. R&D programs

3. Technical information handling

Accordingly, the picture of each of these areas is developed fully in one of the next three chapters, in this order:

Ch. VI. Guide Through the R&D Center's Organization

Ch. VII. Guide Through the R&D Center's R&D programs

Ch. VIII. Guide Through the R&D Center's Technical Information Activities

These separate pictures of R&D center activities are combined in a single practical example of the overall guide for one major Government laboratory, which is presented in the last chapter, viz:

Ch. IX. Case History - Guide Through the National Bureau of Standards

REFERENCES

(1) Advanced Instrumentation For Nuclear Monitoring. G. Armantrout, A. McGibbon, S. Swierkowski, J. Sherohman and J. Yee. PROCEEDINGS, SYMPOSIUM ON SAFEGUARDING OF NUCLEAR MATERIALS, Vienna, Austria, Oct. 20-24, 1975, UCRL Report 76693. Available from NTIS. Part I Radiation Detectors. Part II Gamma Spectrometer Systems.

(2) A very Portable Analyzer, NUCLEAR NEWS, June 1975, pp. 111-112.

Ch. VI

GUIDE THROUGH THE R&D CENTER'S ORGANIZATION

Outline

	Page
Organization	47
Organization Chart	47
Changing Nature of the Work	49
The Matrix Plan	50
Support Functions	51
Summary	52

Ch. VI

GUIDE THROUGH THE R&D CENTER'S ORGANIZATION

Organization

The basic picture on how an R&D center is organized--what its major activities are, what functions they perform, and how they interact with one another--is commonly provided in the Center's annual report.

For example, the Lawrence Berkeley Laboratory (LBL) issued an annual report in 1977 which includes six divisions(1). Of these one is on the Laboratory, which discusses LBL's history, organization, staff, programs and facilities. This is followed by a chapter on each of the Laboratory's major divisions: Energy and Environment; Physics, Computer Sciences and Mathematics; Materials and Molecular Research; Biology and Medicine; Chemical Biodynamics; Nuclear Science; and Accelerators. Each chapter lists the division's staff, programs, facility developments and technical accomplishments. Another good example of this is provided in the annual report for 1977 of the National Bureau of Standards (2).

However, to provide a guide through the R&D center's maze that will enable the outsider to reach the sources of technical information he needs to effect a technology transfer, the picture of the R&D center's organization provided by its annual report is not enough, mainly because not many of the Government R&D centers produce such annual reports, and those that are produced may be as much as a year late. Moreover, not all such reports treat the R&D center's organization with the completeness and thoroughness it received in the LBL report cited above.

Nevertheless, a good picture of how the R&D center is organized, that puts into sharp focus each of the principal activities of an R&D center and how they interact on one another, is provided for practically all R&D centers. This is the center's organization plan, commonly spoken of as its "organization chart," that is drawn up in practically every Government R&D center by its management staff. They produce it mainly for their own use, so it is not widely distributed or publicized.

<u>The Organization Chart</u>. Normally the organization chart provides the map that points the way through a large

scale Government activity such as a national laboratory. Each large Government R&D center usually does have such an organization chart, but it is usually of little help in finding your way through the maze that this center presents to the outsider. This is so mainly because the organization chart does remain unchanged over considerable periods of time, but the organization it represents does not.

The organization chart usually presents the major functional divisions of the R&D center and these usually remain unchanged for years until a major "reorganization" takes place in the R&D center as a whole, or in its major departments. Such reorganizations are usually discussed fully in the R&D center's house organs, such as the employee newspaper, so are easy to get information on. But over the years, although the organization chart remains the same, changes go on continuously within each major unit shown on the organization chart. These are incremental, inconspicuous changes, but they go on continually and they add up over the years. The result is that the groups which produce the technological developments in an R&D center are organized in informal ways which are difficult to trace or follow. The directions in which such informal organizations grow are determined by the following factors, which are treated in detail below.*

---

*However, before taking up those treatments consider the following striking example of the informal organization patterns that develop spontaneously in R&D centers.

<u>Example, LLL</u>: At the huge Lawrence Livermore Laboratory there is an Electronics Engineering Department with a staff of about 1,000 personnel. This huge department is nominally supposed to provide, build, install and maintain all electrical power facilities as well as electronic equipment. However, LLL also has a Plant Engineering Department which also provides, builds, installs, electrical power equipment. After many years of wrangling it was decided that Plant Engineering would take care of the substations and all other electrical power equipment between the incoming electrical utility lines and any research buildings in which experimental or test facilities are located. But any electrical power equipment within the research building would be the province of the Electronics Engineering Department.

So, when LLL made a technological development that would be valuable to private industry--a 300 kw dc power supply that provided highly regulated current output as well as the usual highly regulated voltage output--where would you find the source for information on it? In the Electronics Engineering Department--simply because this substation-size unit was located <u>inside</u> a research building!

1. Changing nature of the work
2. The matrix plan
3. Support functions

Changing Nature of the Work. The nature of the work done in R&D centers changes with time. Often the changes are rapid and dramatic. This requires internal shifts in selection and grouping of personnel even within a single major unit of the R&D center. For example, when the testing of atomic weapons was shifted from tests in the atmosphere to underground testing, the same divisions and departments at LLL carried out the tests, but their work became more involved in seismic phenomena and the instruments to detect and measure them, instead of in the outdoor phenomena of blast and rapid temperature rise. But these are internal changes within the existing, established divisions of the R&D center. The only way you can know about them is to grow up with them.

Another good example occurred in the Electronics Engineering Department of LLL. In the vacuum tube era of electronics technology this Department had large fabrication shops in which electronics technicians skilled in assembly work mounted discrete circuit components on metal chasses. The transistor did away with all that and nowadays electronics technicians still assemble complex electronics systems that meet the needs of the engineers and scientists carrying out the R&D center's research activities. But they do that by mounting little plastic, block-shaped solid state devices on printed circuit boards. This requires new printed circuit board fabrication techniques and equipment. So the outsider interested in transferring, to his firm's commercial needs, the field radiation measurement instrument discussed above (Ch. V) would have to find his way through the LLL's maze to find, first, the people in the EE Department who design its pulse height analyzers, and learn from them who, in the shops, fabricates their designs. For this, also, there are no formal lines of organization to follow. The changes are made as the need arises, and they are made gradually and informally. The only persons who know what has happened are those who grew up with it.

Thus, to be able to provide a guide through the maze of an R&D center that the outsider can follow, so he can reach the sources of the technical information required to effect the transfer of a particular technological development, the writer/editor producing such a guide must obtain a good perspective on how the work of the R&D center is changing. He cannot obtain it from the formal patterns of organization drawn by the management staff of the R&D center. He must get it from the people involved.

The Matrix Plan. In most major R&D centers the major divisions of the organization are of two types. One type is based on the research program to be carried out; the other type is based on the technical specialty to be performed. Thus LLL has one major division based on the laser-fusion R&D program, and another based on the magnetic confinement fusion program. But it also has major divisions based on the technical specialty to be performed, such as the chemistry department or the electronics engineering department.

For these different types of major divisions to work together effectively it is universal practice to assign a group of specialists from the technical specialty division to work on a particular project in an R&D division. They are thus a semi-permanent project task force working in an R&D division but on a long-term assignment from the technical function division to which they belong.

Thus, at LLL, specialists on capacitor banks from the EE Department are assigned to work in the Laser-Fusion Research Department to build the capacitor banks which provide the energy to power the huge lasers being developed for the laser-fusion R&D program. Similarly, a group of computer specialists from the EE Department are also loaned to the laser-fusion department to develop the computer control system for operating the laser-fusion department's giant lasers. Likewise, a group of power system specialists from the EE Department are on long-term assignment to the Magnetically-Confined Fusion R&D Program Department to develop the power supplies needed to operate the high energy neutral beam sources needed in that program.

Thus, to reach the sources for the technical information needed to transfer technology developments in the field of capacitor banks or of high energy power supplies, it would be necessary to guide the outsider to contact the EE Department and find out how and where their specialists on these technologies have been assigned to work, rather than routing them to the R&D divisions for which those technology developments were created.

This cross-connection arrangement for assigning workers in a technical specialty department to work in an R&D program department is known as the "matrix plan." Where it is used extensively, providing a guide for the outsider to reach those who produced the technology he wants to transfer can be a headache. But once such a guide is provided, it can be enormously helpful in promoting the technology transfer process.

Support Functions. Probably the area of greatest concern, in being able to track down the sources of technical information within the R&D organization, is not the pattern of the agency's structure shown on the organization chart but the informal, though basic factor, of how its technical support functions are organized. Does the agency have its own in-house electrical/electronics and mechanical engineering departments, or does it contract out its engineering support activities to outside firms as the need arises?

Some R&D agencies, like the Lawrence Livermore Laboratory, provide all technical support services by in-house engineering departments--the electrical/electronics (EE) department for the creation of all electrical and electronic technological developments; the mechanical engineering (ME) department to create all mechanical and structural technological developments. And each of these departments employs over 1,000 people. The Los Alamos Laboratory, also engaged in atomic weapons development, however, provides the bulk of its engineering support services by farming the work out to private contractors.

The organization pattern which provides technical support services by in-house departments means the agency has to support a large permanent, technical organization. Those agencies which follow the outside contracting approach are more flexible and need a smaller permanent staff. But the outsider seeking the information needed for technology transfer is not concerned about which pattern of organization costs less or works best. What he wants to know is: Which pattern does the agency follow? Because the road to finding technical information is radically different in the two organizational structures.

If the agency provides engineering support services by in-house departments, it is easy to track down the information on a particular development because you can find the man who created it within the agency. If the engineering was done by outside contractors, you first have to find out who the contractor was, then you can discuss any problems you want the solutions for with their staff. However, the appropriate members of their staff may be very difficult to track down because the agency for which the work was done has no control over, and no records of, the contractor's personnel.

Summary

In short, there is no easy way to find a guide through the maze of an R&D center's organization. The center's annual report is of some help, but those reports are few and far between, they are often late in coming out, and they rarely give the full story.

Another helpful source for developing a guide through the maze of an R&D center's organization should be the R&D center's organization chart, but these charts are developed by the center's management staff for the management staff, and show only the center's formal organization pattern. The informal pattern is dictated by the organizational changes from day to day that are determined by the changing nature of the work, by the matrix plan, and by whether support functions are provided by internal departments or by outside contractors. It is this informal organization pattern that determines how work gets done in the R&D center, and by whom. The only way to know this informal organization pattern is to know the people involved in its formation or who grew up with it.

REFERENCES

(1) A REPORT ON THE LAWRENCE BERKELEY LABORATORY, Pub. No. 209/July 1977, Public Information Dept., Lawrence Berkeley Laboratory, University of California, Berkeley, CA, July 1977, 77 pp.

(2) NATIONAL BUREAU OF STANDARDS: A National Resource For Science and Technology, Fiscal Year 1978. Nat. Bur. Stand. (U.S.) Spec. Publ. 538, 86 pages (July 1979). For sale by Supt. of Documents, U.S. Government Printing Office, Washington, D.C. 20402. Stock No. 003-003-02093-1.

(3) 1980 LBL OVERVIEW. Pub. 296/Feb. 1980/Technical Information Dept./12,500 M, Lawrence Berkeley Laboratory, University of California, Berkeley, CA.

Consult Also:

(4) LAWRENCE RADIATION LABORATORY, Berkeley, 1967. Pub. No. 61/3.5 M/Aug. 1968 by TECHNICAL INFORMATION DIVISION, LBL Berkeley. University of California, Berkeley, CA, 1967, 124 pp. Contains comprehensive review articles on R&D programs and section on spin-offs.

(5) BROOKHAVEN NATIONAL LABORATORY Annual Report, July 1969. BNL 50169(AS-23), 243 pp. Available from NTIS.

(6) ARGONNE NATIONAL LABORATORY 1968 Annual Report, 92 pp. Argonne National Laboratory, 9700 S. Cass Ave., Argonne, ILL 60439.

(7) 1968 AT THE OAK RIDGE NATIONAL LABORATORY. 1968, 88 pp. Oak Ridge National Laboratory, Oak Ridge, Tenn. Presents comprehensive review articles on its R&D programs.

Ch. VII

GUIDE THROUGH THE R&D CENTER'S R&D PROGRAMS

Outline

	Page
R&D Programs	55
Roots	56
Current Activities	58
Local Initiative	60
Summary	61

Appendices

App. 7-1	Data, Applied Physics Laboratory	119
App. 7-2	Data, Jet Propulsion Laboratory	121
App. 7-3	Data, Oak Ridge National Laboratory	123

Ch. VII

GUIDE THROUGH THE R&D CENTER'S R&D PROGRAMS

R&D Programs

To be effective in transferring technology from a Government R&D center to private industry, the outsider must not only be able to find his way through the maze of the center's organization, he must also be able to find his way through the maze of the center's accumulation of research and development (R&D) programs.

For example, a valuable technology development for industry created in the National Bureau of Standards is a satellite-controlled, computer-based digital electronic clock. The outsider who wants to transfer this development to meet his firm's commercial needs would have to know that it was developed in the National Bureau of Standards' R&D program on Time and Frequency Standards. This program of NBS is concentrated in its branch at Boulder, Colorado. So knowledge of the R&D program, in which the technological development he wants to transfer is being worked on, would lead the outsider to seek the technical information needed to effect its transfer in the Boulder, Colorado laboratories of NBS, not in its main laboratories at Gaithersburg, Maryland.

Similarly, a valuable technological development produced at LLL is a computer mechanism for converting the different shades of darkness of a black-and-white X-ray photo into bands of different colors, thus making the differences in density of the X-ray film much sharper in contrast and hence much easier to detect (1). This technological development would be very valuable for industry to use in the nondestructive testing of heavy metal machine parts such as forgings and castings. For the outsider to transfer this technological development to commercial application he would have to know that it was worked on in the Flash X-ray Test activity of LLL's R&D program on atomic weapons. Hence he would have to search in the activities of that program to find the people who could give him the technical information needed to effect transfer of that development to commercial application.

Thus, to find his way through an R&D center so as to find the information needed to effect a technology transfer, the outsider must have a clear picture of the R&D programs being worked on at that center.

The key to developing a clear picture of the R&D programs carried on at an R&D center is to look at its accumulation of R&D programs historically. We will therefore discuss the programs at R&D centers on a historical basis. That is, what R&D programs did the center start with (discussed in Roots); what programs does it now have (discussed in Current Activities); and what R&D programs is it working into for the future (discussed under Local Initiative).

Roots. To provide a guide through the maze of a research center's R&D programs one must start at its roots; that is, where the center began. The facilities established for a major laboratory, when it is started, naturally determine the direction of its initial R&D activities. Also, quite naturally, the original facilities at the center are added to and expanded over the years, so its R&D programs accumulate in the same direction as those adopted at the beginning.

For example, the Lawrence Berkeley Laboratory started with the construction of the first high energy particle accelerator for research into the basic structure of matter --the cyclotron--invented by E. O. Lawrence (2). LBL went on to the building of a string of successively more powerful accelerators which culminated in an accelerator capable of multi-billion electron volt outputs--the Bevatron. Consequently, the R&D programs that accumulated at LBL were in the same direction as the earliest ones--that is, toward further and further penetration into the basic structure of matter. Thus the accumulation of R&D programs that occurs at a major R&D center is a framework built on its earliest programs as a foundation.

To provide a guide through the maze of a center's R&D programs one must therefore find a good source of information on its earliest programs. Fortunately, the great majority of our national laboratories were all born at about the same time--during the period of World War II. Also, fortunately, there is a good directory that gives brief summaries of the initial resources of all these R&D centers --their staffs, facilities, funding, etc.--and their initial R&D activities (programs). Called CONTRACT RESEARCH AND DEVELOPMENT ADJUNCTS OF FEDERAL AGENCIES: AN EXPLORATORY STUDY OF FORTY ORGANIZATIONS, it lists all forty of these organizations (Fig. 7-1). These are primary, interdisciplinary laboratories which have been doing the bulk of DoD*, AEC* and NASA* work, hence are the greatest fountains of Government-produced technology to be found (3).

---

*See Table of Acronyms, p. IV.

TABLE I-3. DATE ESTABLISHED, R & D BUDGET, TOTAL STAFF, PRINCIPAL LOCATION, AND PARENT ORGANIZATION, PANEL OF 40 ORGANIZATIONS MEETING THE STUDY SELECTION CRITERIA, FISCAL YEAR 1967

Fig. 7-1 – Summary of data on 40 major Federal contract R&D laboratories

	Date Established[a]	R & D Budget (Millions)	Total[b] Staff	Principal Location	Parent Organization
ATOMIC ENERGY COMMISSION					
Ames Laboratory	1942	8.3	1,012	Ames, Iowa	Iowa State University of Science and Technology
Argonne National Laboratory	1942	75.3	5,388	Argonne, Illinois	University of Chicago
Bettis Atomic Power Laboratory	1948	68.3	2,845	Pittsburgh, Pennsylvania	Westinghouse Electric Corporation
Brookhaven National Laboratory	1947	48.3	3,376	Upton, Long Island, New York	Associated Universitites, Inc.
Cambridge Electron Accelerator	1956	3.6	193	Cambridge, Massachusetts	Massachusetts Institute of Technology and Harvard University
Knolls Atomic Power Laboratory	1946[a]	43.7	2,080	Schenectady, New York	General Electric Company
Lawrence Radiation Laboratory, Berkeley	1936	37.7	3,465	Berkeley, California	University of California
Lawrence Radiation Laboratory, Livermore	1952	106.2	5,534	Livermore, California	University of California
Los Alamos Scientific Laboratory	1943	91.6	4,229	Los Alamos, New Mexico	University of California
Mound Laboratory	1943	12.9	2,068	Miamisburg, Ohic	Monsanto Research Corporation
Oak Ridge National Laboratory	1943	79.2	5,006	Oak Ridge, Tennessee	Union Carbide Nuclear Company
Pacific Northwest Laboratory	1942[a]	40.4	2,458	Richland, Washington	Battelle Memorial Institute
Princeton-Pennsylvania Accelerator	1956	5.1	315	Princeton, New Jersey	Princeton University and University of Pennsylvania
Princeton Plasma Physics Laboratory	1951	6.3	423	Princeton, New Jersey	Princeton University
Sandia Laboratory	1945[a]	108.6	8,206	Albuquerque, New Mexico	Western Electric Company
Savannah River Laboratory	1950	16.3	791	Aiken, South Carolina	E. I. du Pont de Nemours & Company
Stanford Linear Accelerator Center	1961	22.4	1,228	Stanford, California	Stanford University
OFFICE OF THE SECRETARY OF DEFENSE					
Logistics Management Institute	1961	1.2	43	Washington, D.C.	Not-for-profit corporation
Institute for Defense Analyses	1956	14.9	625	Arlington, Virginia	Not-for-profit corporation
DEPARTMENT OF THE AIR FORCE					
Aerospace Corporation	1954[a]	74.8	3,970	El Segundo, California	Not-for-profit corporation
Analytic Services, Inc.	1958	1.3	84	Falls Church, Virginia	Not-for-profit corporation
Electromagnetic Compatibility Analysis Center	1961	4.7	294	Annapolis, Maryland	IIT Research Institute
Francis Bitter National Magnet Laboratory	1960	2.5	109	Cambridge, Massachusetts	Massachusetts Institute of Technologv
Lincoln Laboratory	1951[a]	74.2	1,840	Lexington, Massachusetts	Massachusetts Institute of Technology
MITRE Corporation	1958	36.4	2,050	Bedford, Massachusetts	Not-tor-profit corporation
RAND Corporation	1948[a]	22.8	1,125	Santa Monica, California	Not-for-profit corporation
DEPARTMENT OF THE ARMY					
Center for Research in Social Systems	1956	2.1	131	Washington, C.C.	The American University
Human Resources Research Office	1951	3.4	264	Alexandria, Virginia	The George Washington University
Research Analysis Corporation	1948[a]	11.8	643	McLean, Virginia	Not-for-profit corporation
DEPARTMENT OF THE NAVY					
Applied Physics Laboratory (Johns Hopkins Univ.)	1942	62.9	2,564	Howard County, Maryland	The Johns Hopkins University
Applied Physics Laboratory (Univ. of Washington)	1943	2.9	170	Seattle, Washington	University of Washington
Center for Naval Analyses	1942[a]	9.3	450	Arlington, Virginia	University of Rochester
Hudson Laboratories	1951	4.7	381	Dobbs Ferry, New York	Columbia University
Naval Biological Laboratory	1946[a]	2.0	148	Oakland, California	University of California
Ordnance Research Laboratory	1945[a]	8.1	468	State College, Pennsylvania	The Pennsylvania State University
NATIONAL AERONAUTICS & SPACE ADMIN.					
Bellcomm, Inc.	1962	10.1	445	Washington, D.C.	AT&T and Western Electric Company
Jet Propulsion Laboratory	1939	226.3	4,586	Pasadena, California	California Institute of Technology
NATIONAL SCIENCE FOUNDATION					
Kitt Peak National Observatory	1958	12.4	243	Kitt Peak, Arizona	Association of Universities for Research in Astronomy, Inc.
National Center for Atmospheric Research	1960[a]	14.3	463	Boulder, Colorado	University Corporation for Atmospheric Research
National Radio Astronomy Observatory	1956	5.2	212	Green Bank, West Virginia	Associated Universities, Incorporated
TOTALS		$1,382.5	69,925		

a. When centers were preceeded by existing organizations, meaningful establishment dates for the purpose of this tabulation were difficult to determine. In these cases, dates were selected based on when the organization began to assume center characteristics and missions.

b. Budget and manpower data supplied by the individual organizations usually included work performed for all sponsors - not just the primary sponsor - where multiple sponsorship was present. This, plus the differences in accounting procedures and fiscal year dates, cause these data to vary somewhat from other compilations.

In a very compact, tabular form, this directory tells you the history of each laboratory, its resources in facilities and manpower, which technical areas it specializes in, and what are its major technological achievements. Good examples are the data on the Johns Hopkins Applied Physics Laboratory (Appendix 7-1), a well of technology involved in Naval weapons and NASA satellites; Jet Propulsion Laboratory (Appendix 7-2), a tremendous concentration point for the technology of spacecraft and the deep space communication systems; and Oak Ridge National Laboratory (Appendix 7-3), a treasure house of technology on nuclear reactors, desalination, and radioactive isotopes.*

By consulting this directory one can get a good foothold on the job of preparing the guide to the R&D programs at a particular R&D center.

Current Activities. To find your way through an R&D center to the information you need to effect technology transfer, one must obviously know what R&D programs are being worked on in that agency in the present, as well as knowing what programs it started with in the past. To determine these current programs is a simple matter, in the newer agencies, because Congress usually issues a clear mandate to new R&D agencies as to what R&D areas they are to work in. But as the years go by each such agency picks up R&D projects that are only marginally related to its main area of activity. This is especially true when the agency's main program tapers off and gets reduced Government funding. The R&D agencies then scramble to get work from other agencies, or to develop new R&D programs for which Congress can be persuaded to provide funds. For example, the space R&D agency (NASA) is now doing a considerable amount of work for the Department of Energy on the development of solar energy systems. Likewise, the Argonne National Laboratory, originally primarily a nuclear fission power plant-development laboratory, is now heavily involved also in the development of electrical storage batteries.

There is no systematic way and no single source for finding out what additional R&D programs an R&D center has picked up over the years. However, there is one approach that is very helpful: Study the list of reports and publications that the center, whose current R&D programs you want to determine, has issued or published. In some centers such a list is published and is usually

---

*These laboratories have other areas of specialization besides those mentioned in this paragraph. Consult the appendices listed, also these laboratories' annual reports.

available from that center's technical library or technical information center. Even where such a list is not published it is almost always compiled by the center's technical library and a Xerox copy may be obtained by an outsider on request.

To study these report and publication lists is obviously a chore, but doing so can be especially rewarding because it will readily reveal any new R&D programs that a particular center would not be expected, on the basis of its past activities, to be carrying on. Thus, for example, the LLL is basically an atomic weapons R&D center, so one would not expect it to be engaged in R&D programs that offer valuable technology developments for transfer in the geothermal energy field. Yet even a brief perusal of its recent publication list discloses a report entitled: The LLL GEOTHERMAL ENERGY PROGRAM STATUS REPORT, JAN. 1976-JAN.1977 (4). Its value for technology transfer in the geothermal energy field is indicated by the following "Preface" from that report.

PREFACE

The Lawrence Livermore Laboratory's Geothermal Energy Development Program, funded by the U.S. Energy Research and Development Administration, was initiated in January 1974. This report, which is the third major status report, describes significant events and detailed results obtained from January 1976 through January 1977. The timing is purposely chosen to coincide with the completion of several significant laboratory and field experiments. Furthermore, this report serves as documentation of the material presented at a major program review meeting held at Livermore on December 14-16, 1976, and attended by more than 80 industry representatives. The purpose of the meeting was (1) to provide detailed current information on the LLL program content, technical results, and goals; (2) to seek participation from industry in helping review program direction; and (3) to encourage greater industry participation in future work toward the commercialization of our results. It is hoped that this report will further these objectives.

The authors or contributors are identified in the report and in the tables of contents. These are the people who have done the detailed technical work and should be recognized. The editors have produced introductory material for each section and for the report as a whole.

A. L. Austin

Where a study of a major laboratory's report and publication lists is not feasible, a partial measure for uncovering its R&D programs is to be found in its annual report. However, this has all the same shortcomings as the use of the annual report of an R&D center to determine the center's organization that were cited above (Ch. VI - Organization). That is, not many R&D centers issue annual reports; those that are issued may be late, and they may not be very thorough or complete on this subject.

Local Initiative. In constructing the guide through the accumulation of R&D programs at a major laboratory, it is of course vital to consider both its R&D program accumulation from the past (Roots) and those programs it is currently engaged in (Current Activities). But it is also possible and advisable to include a treatment of those programs that the R&D center will be working on in the future.

R&D centers get their R&D programs from three major sources. Some are imposed by law. For example, NASA was mandated by Congress to put a man on the moon; and the National Bureau of Standards is required by act of Congress to do R&D towards developing standards in certain commercial fields, like fireproof materials.

The great majority of their programs, however, are acquired by R&D centers by contract. Thus the LLL is assigned R&D programs for developing atomic weapons, through contract from the Energy Department. In some cases NBS even carries out R&D programs, such as one for developing new materials to be used in dentistry, through contracts with trade associations.

But in the course of carrying out their programs the R&D centers build up know-how and facilities that point to problems they could solve in related fields. The center then starts up an R&D program in such a related field on its own initiative. After such programs have shown some promise, the R&D center asks for funding for it from the parent agency, which funds it usually gets. These programs thus rise out of local initiative at the R&D centers.

For example, the LLL has huge resources and much activity in developing fusion power through use of lasers. It was quite natural for the scientists at LLL to think of using their high power lasers to accomplish isotope separation by use of laser energy, so they started some research in this related field on their own initiative. Soon the results became promising and they obtained funding from the

Energy Department to make isotope separation by lasers a full scale R&D program of the Laboratory (5).

An even greater example occurred in the space program when NASA scientists found that their earth-orbiting satellites, when equipped with radiation sensors and cameras, could provide pictures of the earth that were very helpful in detecting mineral-bearing geological formations, in determining crop yields, in making ecological surveys, etc. NASA has therefore started its Landsat satellite program on its own initiative and is now receiving funding for it from Congress.*

These "local initiative" R&D programs seldom make up more than a small part of an R&D center's accumulation of R&D programs, but they often present the possibility of a rich source of transferable technology for the future, so they should be studied and included in the guide to be produced on an R&D center's accumulation of R&D programs.

Summary

Major laboratories (R&D centers), in the course of their existence, accumulate an assortment of R&D programs.

---

*Other examples of such "local-initiative" R&D programs are the following. Los Alamos Scientific Laboratory (LASL) initiated a program for drilling a hole deep into the earth and pumping water down into it until it reached the hot rock formations deep in the earth. There the water would be turned into steam which would escape, through another hole drilled deep in the earth, to the surface where it would drive turbines to generate electrical power.

Similarly, at LLL, they studied the current practice of generating geothermal energy. It is standard practice to take the stream of hot water mixed with steam that gushes up from the geothermal wells, separate off the steam and use it to drive turbines for power generation. This wastes the energy in the hot water which gushes up, at high velocity, along with the steam. So the LLL scientists proposed projecting the stream of both water and steam against a Pelton wheel, thus extracting the energy from both the steam and the hot water to drive an electrical generator. This has since become a regular R&D program at the LLL funded by the parent agency, the Energy Department (4).

A guide through the maze of this program accumulation can be developed by studying the accumulation on a historical basis. The programs which the center starts out with are largely determined by the nature of its initial R&D facilities. As the laboratory grows and expands, it develops facilities and programs closely related to those it started with. The center's current R&D activities are thus a framework built on its initial facilities and programs as a foundation.

In the course of these earlier and its current programs the center develops capabilities for solving R&D problems in other fields. It starts work on these in a small way, on its own initiative. When it reaches a promising level of activity in these programs it gets funding for them from the parent agency. Thus a center's future R&D programs are those that evolve from its local initiative in reaching into other areas of R&D.

REFERENCES

(1) Digital Image Processing: A Flexible Diagnostic Tool, ENERGY AND TECHNOLOGY REVIEW, LLL Publication UCRL 52000-75-12, Dec. 1975, pp. 15-21. Obtainable from NTIS.

(2) 200 Man-Years of Life - The Story of Ernest Orlando Lawrence. Daniel Wilkes. THE PHYSICS TEACHER, Vol. 3, No. 6, Sept. 1965, pp. 1-10.

(3) CONTRACT RESEARCH AND DEVELOPMENT ADJUNCTS OF FEDERAL AGENCIES: AN EXPLORATORY STUDY OF FORTY ORGANIZATIONS. Prepared by Denver Research Institute, University of Denver, for the National Science Foundation, March 1969. Obtainable from Denver Research Institute, University of Denver, Denver, Colorado

(4) THE LLL GEOTHERMAL ENERGY PROGRAM STATUS REPORT - JAN 1977. UCRL Report 50046-76. April 27, 1977, 173 pages. Available from NTIS, Springfield, VA 22161.

(5) Laser Isotope Separation - Research and Development. Ch. 12, pp. 483-508, LASER PROGRAM ANNUAL REPORT - 1974. Lawrence Livermore Laboratory, UCRL 50021-74, March 1975. Obtainable from NTIS, Springfield, VA 22161.

Ch. VIII

GUIDE THROUGH THE R&D CENTER'S TECHNICAL INFORMATION ACTIVITIES

Outline

	Page
Technical Information Handling	65
Information Generation	65
Contractual Reports	66
Internal Reports	66
Drawings	67
Equipment Manuals	68
Specifications	70
Standards Manuals	72
Standards Manual Case History - Microcomputers	73
Information Storage - Contractual Material	76
Contractual Reports	77
Report Storage - R&D Centers	77
Report Storage - Central Repositories	78
Report Storage - Regional Repositories	79
Information Storage - Internal Technical Material	81
Internal Material Storage Maze	81
Internal Material Storage Maze - Case History	81
Internal Material Storage Guide	82
Information Dissemination - Contractual Material	83
Abstract Bulletins	83
Technology Transfer Centers	85
Direct Dissemination	87
Summary	87
Summary - Contractual Reports	87
Generation	87
Storage	87
Dissemination	88

(Cont.)

Outline (Cont.)

	Page
Summary - Internal Technical Information Material	88
Generation; Internal Information Material	88
Internal Reports	88
Drawings	89
Equipment Manuals	89
Specifications	89
Storage: Internal Information Material	90
Dissemination: Internal Information Material	90
Offices of Technology Transfer	90
R&D Center Libraries	91
Guides to Technical Information Handling	91

Ch. VIII

GUIDE THROUGH THE R&D CENTER'S TECHNICAL INFORMATION ACTIVITIES

Technical Information Handling

To achieve a technology transfer economically, the outsider must be able to obtain total technical information about the development to be transferred, so that he does not have to reinvent the wheel. This is difficult to do because the maze he has to go through at the R&D center to get that information is composed of three separate mazes, wrapped one inside another.

The outermost maze is the organization maze. Ways of developing a guide for the outsider through this maze have been given above (Ch. VI). Next is the maze of the R&D program accumulation at the R&D center. A technique for building a guide that will enable the outsider to wend his way through that maze is also presented above (Ch. VII).

The next and innermost maze that the outsider has to work his way through to get the information needed to effect a technology transfer is the maze of the practices followed by the R&D center in handling the technical information produced at that center. To produce a guide through this maze requires consideration of each of the three major areas of technical information handling: Generation of the technical information, its storage, and its dissemination.

Information Generation

It is almost a universal assumption that the technical information generated in R&D centers consists entirely of the R&D reports issued by the center, which reports are disseminated to the public, mainly through the National Technical Information Service.* This is very far from the truth.

The technical reports issued by Government R&D centers or agencies, and which the public generally considers as the main source of technical information useful in effecting technology transfer, constitute only a fragment of the technical information generated in Government R&D centers.

---

*Only unclassified R&D reports are disseminated. Classified reports are sent only to parties holding appropriate clearances and who are authorized to receive them.

The total output of technical information in those centers comprises contractual reports, internal reports, drawings, specifications, and equipment manuals. All these different types of documents contain technical information that is needed for effecting transfer of a technological development, and the information in all of them must be gathered and integrated to effect a technology transfer economically --without reinventing the wheel. The guide through the technical information handling maze of the R&D center must therefore make clear how all these different types of technical information sources are generated. Such a clarifying picture follows.

Contractual Reports. On all R&D projects funded by the Federal Government, the final step required is the submission of an R&D report which tells what was accomplished and how it was achieved. Such a report is normally considered a part of the contractual requirement for carrying out the R&D project--i.e., a part of what was paid for by the funding for the project.

Since many R&D projects run on for years, these reports are often submitted as annual progress reports. And because they are used as the basis for evaluating the performance on projects funded in the past and for funding projects in the future, these reports are often professionally done, produced by professional writer/editors working in close cooperation with the top scientists and engineers involved in the R&D project. A good example is the report on the Geothermal Energy R&D project at LLL for the year 1976 noted above (Ch. VII, Ref. (4)).

But these contractual reports are merely the tip of the iceberg of technical information generated at the R&D center on all its technological developments. The mass of the iceberg, contained below the surface, is composed of the internal reports, the drawings, and the specifications produced on each technological development created in the R&D center.

Internal Reports. Of far greater number than the contractual reports are the internal reports generated on each technological development created at the R&D center. The internal reports are often engineering studies, to find the solutions to problems that crop up in the course of producing a particular technological development. These are considered important technical material, just like drawings, so should be preserved. Consequently, they are often stored in the same filing system which stores the drawings. But a large portion of the internal reports are mainly the periodic progress reports, produced by the engineers who are creating the technological development, to inform their immediate supervisors of the progress being made on the project.

These internal reports can contain a lot of alibis for mistakes made and apologies for not making expected progress or for falling behind schedule. But they often contain one valuable thing. That is, they often show the process of reasoning for particular steps taken and this can be the most helpful thing of all to the outsider striving to convert a particular technological development from a Government R&D program into a successful commercial product.

A very good example of this is afforded in the article cited above (Ch. II, Ref. (3)) on avalanche transistor circuits. This highly successful article was in effect the result of stitching together a few EE Department internal reports in which the reasoning behind such circuits was developed.

Thus, the internal reports, which are produced in great volume in each R&D center but are not circulated outside it, often contain the vital technical content necessary for effecting a technological transfer. The guide through the maze of technological information handling at an R&D center should therefore give a clear picture of where and how the outsider can find out which of these internal reports have been produced, and how he can get access to their contents.

Drawings. The drawings produced to achieve a particular technological development of course contain details of manufacture and construction that can save the person, who is planning to effect a transfer of that development, a great deal of time and money. A complete set of drawings is usually produced in the drafting section of the engineering department working on that development. But this section may be split up into small groups that are assigned, for a long term, to a particular R&D project. So where their drawings are stored becomes a question of how the R&D center is organized. Fortunately, in most cases the drawings produced by all drafting groups are stored in one place because it is necessary to make blueprint copies of them or to reduce them to microfilm or microfiche size to reduce storage space. The blueprint production or microfilming facility then becomes a central unit which stores the drawings for all groups. Hence, to find where the drawings produced in an R&D center are stored, one must ask the manager or supervisor of the blueprint or microfilming facility.

In different engineering departments the drawings are produced and numbered according to different procedures. The guide through the maze of an R&D center's information handling procedures should therefore treat these in detail for each department.

The system for numbering and storing drawings used in the Electronics Engineering Department at LLL is a unique one that is very helpful in effecting technology transfer. Instead of giving each drawing a unique number, the drawings related to a particular equipment unit all have a common main number, which indicates what equipment unit they belong to. Added to this main number is a small, dash-connected suffix number which identifies the individual drawing within that group. Hence a glance at the drawing's identifying number immediately identifies which equipment unit it belongs to.

The EE Department drawing system also contains this feature that is immensely helpful in effecting the transfer, to commercial use, of a particular technology development produced in that Department.

Equipment Manuals. In the drawing package for each equipment unit one of the low suffix numbers, say the -2, is always reserved for the schematic or assembly drawing. Thus one knows instantly what drawing to look for to find out how the equipment unit operates or is assembled. In addition, another low suffix number of the drawing package is always reserved to identify, not a drawing, but a document that is always to be included in each drawing package. This document is called the "Description and Instruction" document (D&I) which gives a description of how the equipment unit works and instructions for its installation, operation, maintenance, and servicing.

These D&I's are the equivalent of the equipment manual usually produced by the armed services for each piece of equipment the soldier or sailor uses. They are thus immensely valuable for technology transfer because, if you know how the equipment operates and how it should be serviced, you have the bulk of the technical information needed to effect its conversion to a commercial product. For example, LLL designed a raster type oscilloscope for precision measurement of microsecond time intervals (see Fig. 4-2, Ch. IV). The D&I prepared for this oscilloscope (Fig. 8-1) enabled other AEC (Dept. of Energy) contractors to build and use that oscilloscope (1)(2)(3).

This is an extraordinarily useful arrangement for facilitating technology transfer. Unfortunately, there is

L-300 RASTER OSCILLOSCOPE OPERATION AND MAINTENANCE MANUAL

TABLE OF CONTENTS

		Page
1.0	GENERAL DESCRIPTION	5
2.0	SPECIFICATIONS	7
3.0	OPTIONS AND ACCESSORIES	13
4.0	OPERATING INSTRUCTIONS	17
5.0	THEORY OF OPERATION	25
	Document Numbering System	25
	Generation of Display	28
	Main Frame -001	31
	Common Printed Circuits -200	39
	Standard Timing Format Circuits -300	42
	Fast Timing Format Circuits -400	65
	Slow Timing Format Circuits -500	71
6.0	NEW INSTRUMENT CHECKOUT	84
7.0	CRT REMOVAL/REPLACEMENT	98
8.0	CALIBRATION	101
9.0	MAINTENANCE	107

Fig. 8-1. Table of Contents of D&I on L-300 high speed oscilloscope system indicates the valuable technology transfer information (operation and servicing) provided by D&I's (equipment manuals). (See Ref. (3).)

a huge flaw in it. The workload of producing such a D&I is roughly one man-day per page. So, for a moderate sized D&I of fifty or sixty pages, this means an expenditure of ten or twelve weeks' time. After an engineer has completed a particular technological development he is not willing to put in so huge an outlay of time to prepare a D&I on it, so the engineer forgets the D&I and moves off to other development jobs. As a result, as a check made by the author of this book showed, only a fraction of the drawing packages produced in the EE Department have a D&I document.

Undoubtedly this gross negligence in the preparation of the D&I's or manuals needed for new technology developments occurs in other R&D centers. But the D&I document or manual is the single most helpful item in effecting conversion of a technology development from a Government R&D program into a commercial product. This widespread failure to produce D&I's on the technology developments created in Government R&D programs therefore adds up to a vast loss in valuable new technology to American industry.

Specifications. In practically all R&D centers a specification is drawn up on a technological development before it is built. These specifications give the performance requirements that the system, subsystem, or equipment unit that is to be built shall meet. But a good specification contains, besides performance requirements, certain utility requirements. These are the requirements that the equipment to be built shall meet to make its operation safe, convenient, etc.

For example, in the case of electron microscopes, there is this important utility requirement. When the field coils which focus the electron beam are energized, the current through them heats up the whole microscope structure. Since chemical activity increases rapidly with temperature rise, it is important that this turning on of current in the focusing coils should not heat up the microscope more than a few degrees. Otherwise, chemical changes will occur in the samples which will make the observation of them under the microscope misleading. Hence this requirement was also included in the specification for the LLL microscope (Fig. 8-2(B)). Obtaining the specification on the equipment unit to be transferred can therefore be very helpful in converting it to a successful commercial product.

Normally, each engineering department has a few specialists who write all its specifications. In the EE

6 Jan 64

2.6 Be as versatile as is compatible with highest performance in electron microscopy.

3.0 EQUIPMENT

The microscope shall include these major equipment units or assemblies.

3.1 Power Supply

3.2 Electron gun

3.3 Electron optics - electro-magnetic lens system to consist of five (5) lenses as follows:

3.3.1 condenser lens system

3.3.2 objective lens

3.3.3 intermediate lens

3.3.4 projector lens

3.4 Specimen chambers

3.5 Viewing system

3.6 Photographic system

3.7 Vacuum system

3.8 Accessories

4.0 DESIGN FEATURES AND OPERATING CHARACTERISTICS

Design features and operating characteristics of the major equipment units shall be:

4.1 Power supply

4.1.1 shall provide accelerating voltages for beam of 30, 50, 75, and 100 kv (see Design Options Item 8.1.1.1)

4.1.2 shall be fed from single-phase, 110-volt, 60-cycle AC main (see Design Options Item 8.1.1.2)

4.1.3 power consumption shall not exceed 3.0 to 3.5 kva

4.1.4 shall provide 500- or 1,000-volt AC voltage which can be superimposed on high voltage for voltage-centering the beam.

4.1.5 shall include meters for monitoring the lens currents, beam current, and beam voltage. These meters shall be mounted on the control console.

TITLE ELECTRON MICROSCOPE

(A)

Page 2 of 10

6 Jan 64

5.1.1 ambient room temperature of 68°F to 110°F

5.1.2 normal room air (requiring protection of sensitive relays, etc., from dust and humidity)

5.1.3 normal laboratory vibrations (from pumps, air circulating systems, etc., and other laboratory equipment in adjacent rooms)

6.0 PERFORMANCE

6.1 System as a whole

6.1.1 resolving power for microscopy: 8 Angstroms, guaranteed

6.1.2 resolving index for selected area diffraction: 5×10^{-5}

6.2 Power Supply

6.2.1 accelerating voltage shall be stabilized to 0.001% for 2-minute span

6.2.2 lens currents shall be stabilized as specified below (6.3.3.3, 6.3.4.2, 6.3.5.1)

6.2.3 filament current shall stabilized to 0.1% or better

6.3 Electron optics

6.3.1 overall astigmatism $<0.1\ \mu$

6.3.2 reproducibility of magnification 2%

6.3.3 condenser lens system

6.3.3.1 shall be able to produce beam at specimen (for microscopy) down to 2 micron diameter

6.3.3.2 shall be able to produce beam at specimen (for diffraction) down to 5 micron diameter

6.3.3.3 shall have current stabilized to 0.003% for 2 minutes

6.3.3.4 temperature of coils shall not vary during 3 hours continuous operation, more than 3°C

6.3.4 objective lens system

6.3.4.1 astigmatism shall be such as to comply with 6.3.1

6.3.4.2 current shall be stabilized to 0.001% for 2 minutes

6.3.4.3 temperature rise of coils, during 3-hour continuous operation, shall not be more than 3°C

TITLE ELECTRON MICROSCOPE

(B)

Page 7 of 10

Fig. 8-2. Pages from "modular" system specification on electron microscope (see Ref. (4)). List of system modules (A), and utility requirements (B).

Department of LLL the specifications are each written, instead, by the engineer who is buying the equipment. A "modular" system of preparing specifications (Fig. 8-2A) was prepared in the EE Department, which system uses a fixed format for organizing the material in a specification (5)(6). One can then shift from one specification to another by inserting different modules, within the fixed framework, to present the requirements for different units. This makes it easy for an engineer to learn to write his own specs. Also, by studying a group of specs that have been produced on one type of equipment, he can quickly see what are the important requirements for that type of equipment and can quickly put together appropriate requirement "modules" to compose specifications that meet his need. One person reviews all the specs produced in the Department to see that they all conform to the same format. A computerized permuted index of all specs is then made, so an engineer can quickly find all the previous specs that have been produced on the type of equipment he needs--say on power supplies or computers, or on laser systems.

With this system of producing specifications used by all engineers in the EE Department, it becomes easy to find the specification prepared on the technology development to be transferred. Moreover, it becomes easy to pick out the requirements that will help to convert it to a successful commercial product.*

Standards Manuals. In some R&D centers so much research equipment is produced that it is profitable to follow certain standard practices in their design and construction. These standard practices provide such great savings in time and cost in the design and fabrication of the equipment that they would be immensely beneficial if used in the process of converting a technology development from a Government R&D program into a commercial product.

For example, in the days when atomic weapons tests were done in the atmosphere, it required many trailers full of instrumentation to be built at LLL and shipped overseas to the Pacific Islands where the tests were held. To save time and costs in their design and fabrication, standard instrumentation components and sub-assemblies were developed and supplies of them were stocked at LLL. This permitted great savings in time and cost of both design and fabrication of the equipment.

---

*A recent, comprehensive description of this specification preparation system is presented in Ref. (6).

For instance, so many control rooms are built at LLL for its test and research facilities that it became standard practice to house all control room instrumentation in steel cabinet-like racks formed out of two angle-iron frames, front and rear, bolted to and held together with cross bars. The rack frames, front and rear, are identical, so only one rack frame has to be stocked. Standard cross bars for holding the frames together and standard steel plates for covering the frames to form a cabinet were also produced and stocked. It was thus possible to build, from stored stock, as many control room equipment racks as were required in the test-site trailers or in the test and research facilities built at the Laboratory.

This practice of designing and stocking standard components and sub-assemblies became so extensively adopted at LLL that a STANDARDS MANUAL (7) was produced to cover them. This system has become so useful that it has been adopted widely by other AEC laboratories and it can be installed in each one by merely providing them with a copy of the standards manual. Such a manual, where available, should therefore always be included in the guide for the outsider through the maze of an R&D center's technical information handling practices.

Standards Manual Case History - Microcomputers. A particularly striking example of this standardization, and how it helps to cut time and costs in design and construction and thus would be beneficial in the technology transfer process, is afforded in the standardization practices developed at LLL for microcomputers.

When it became possible to put a major part of a digital computer on a single, tiny, solid state "chip"--say a processor, memory, or input-output device--LLL developed small printed circuit boards, on each of which such a major component of a computer was mounted together with its related circuit elements that enable it to function as a part of the computer. These circuit boards thus constituted modular parts of a computer system (8) that could be plugged into a standard rack to form a functional microcomputer system. To change computer system architecture, it was then necessary merely to change the circuit board modules that were plugged into the rack (Fig. 8-3, 8-4).

These standard circuit-board computer modules and the standard housings, in which were mounted the racks that the modules were plugged into, were manufactured in quantity

---

*LLL pioneered in the development of this modular microcomputer system (8) which has since become universal technological practice.

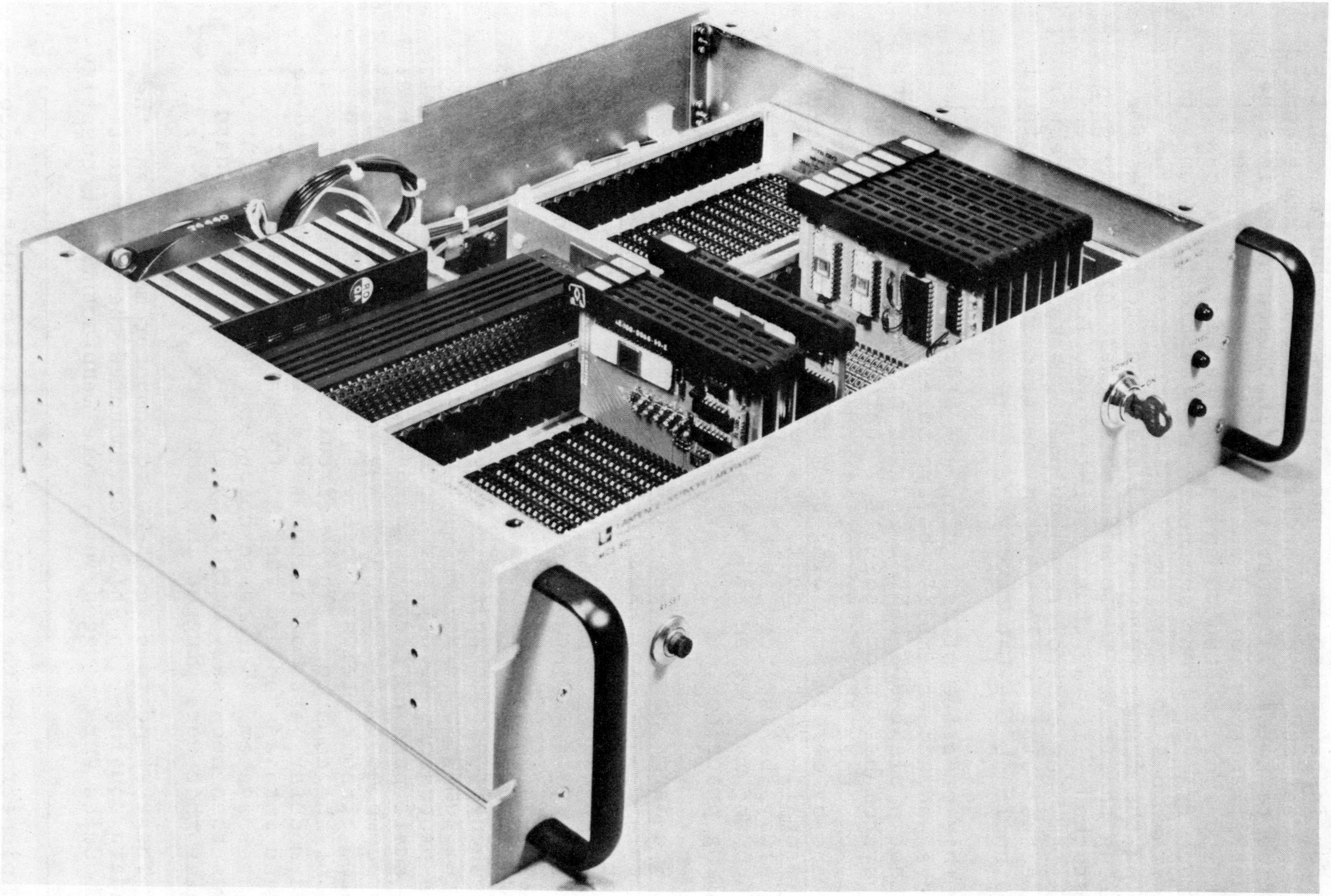

Fig. 8-3. Modular functional divisions of microcomputer system, on standard printed circuit panels, plug into standard rack to form desired computer system.

REFER QUESTIONS TO: EE Components Grp.	LAWRENCE LIVERMORE LABORATORY	ELECTRONICS ENGINEERING **STANDARD** LIVERMORE, CALIFORNIA	UNIVERSITY OF CALIFORNIA	**L E D** 12-80-11-B1
ORIG. K.L. Burris/C.W.J	CLI CARD, MEMORY, PROM, 8-BIT X 2K WORD			DATE 12-1-76
APPROVED [signature]				SHEET 1 OF 4

HIGH-SPEED 2K PROM (450 ns ACCESS), LEA72-9068-32
STOCK NO.: 5975-62643

DESCRIPTION

This board, LEA72-9068-32 uses two Intel 2708 PROM chips, which are 1K X 8. Therefore, these two PROM chips give a total memory of 8_{10} pages when this card is installed in an MCS-80. (With minor modifications the card will provide the same number of memory pages for an MCS-8.) As these PROM chips are "faster" than the 500 ns-state time of the MCS-80, the MCS-80 is allowed to operate at full speed.

SPECIFICATIONS

Speed: Operate the 2708 PROM chips at 450 ns access time

Power: +12 Vdc @ 130 mA

-12 Vdc @ 100 mA

+ 5 Vdc @ 270 mA

IC Specifications

82S42 See EE Standards Manual, Section 11, Microcircuits
2708

LOGIC DIAGRAM

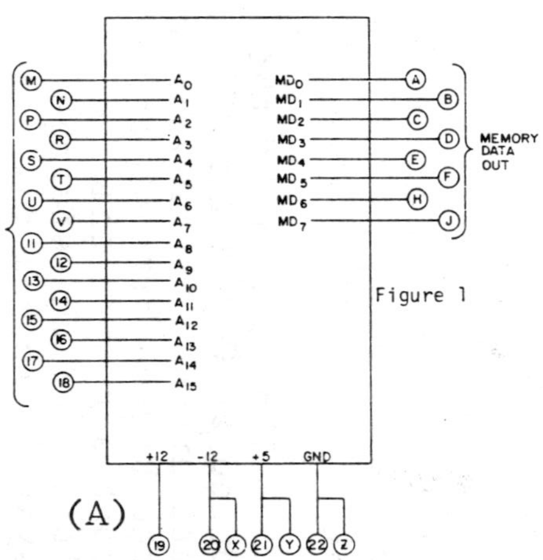

Figure 1

(A)

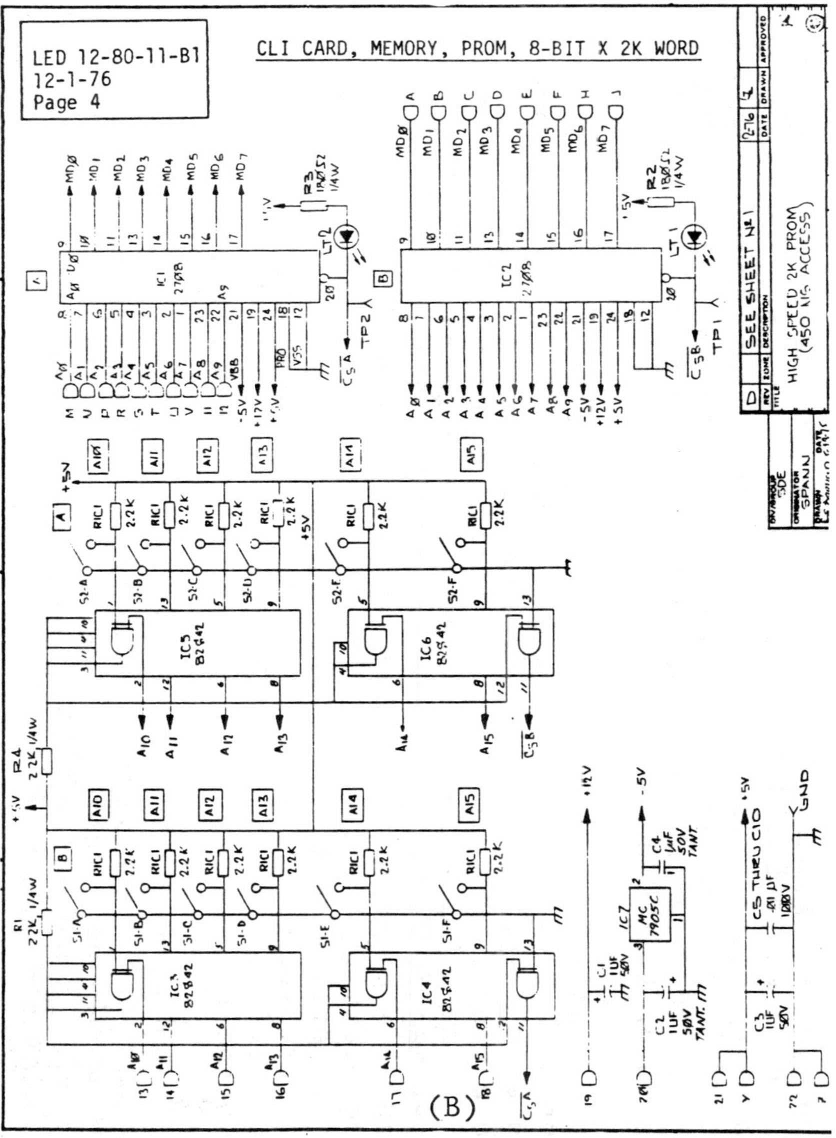

(B)

Fig. 8-4. Data sheets, from Standards Manual, on memory module of modular microcomputer system. Page 1 of Standard (A); page 4 (B).

and stocked at LLL. Hence, when a scientist or engineer at LLL desired to build a microcomputer to control the research machine he was building, he merely drew from stock the standard housing and the standard computer-component circuit boards to plug into it that were required to give him the computer system design he needed (9).

This system is especially valuable, both at LLL and in the technology transfer process, because full information on each computer-system-module circuit board is presented in a single "standard" in the standards manual (Fig. 8-4). By exchange of these data sheets with a commercial firm that manufactures such circuit boards, all the plug-in-module microcomputer technology developed by LLL is made commercially available. Thus, any private firm that wants to benefit from this standardized, modular computer technology development at LLL can do so by obtaining those data sheets from LLL or by purchasing those standard modules and their standard rack housings on the open market (8).

Hence, in producing the guide through the maze of an R&D center's technical information handling procedures, one should be sure to include in it a treatment of the center's standardization practices, as disclosed in its standards manual.

Information Storage - Contractual Material

For the outsider to do an effective job of technology transfer he must not only know of all the different types of technical information material produced at the R&D center, he must also be able to get his hands on them. In short, he must know where that material is stored.

The storage of the technical information material produced at Government R&D centers breaks down into two separate areas. One is the storage of the contractual material--the R&D reports which the center produces to fulfill its contract with the funding agency. The procedures and facilities for the storage of this material are well established, systematic and quite uniform throughout the various R&D agencies. A guide for the outsider to find his way through them can be easily prepared, as described below under Contractual Material.

The other area of storage of technical information material produced in the R&D center is in the storage of the internal documents--the internal reports, the drawings, the specifications and the equipment manuals. It is in these materials that the really rich raw material for effecting technology transfer is to be found. But for this material there is no systematic, universal, storage practice or facility. One just has to piece together the information on the storage of all this internal material by groping around in the R&D center. Some pointers on building a guide for the outsider to cover this area of technical information storage are presented below, following the treatment for the contractual material.

Contractual Reports. When a contractual R&D report is completed, copies of it are not made from a typewritten master on a Xerox machine as needed. Rather, the report is printed and the print run is rarely under a hundred copies. Usually it is around several hundred. This quantity is usually divided as follows: First, a considerable number are distributed to persons, institutions, or Government offices that the author feels would be interested in the report. Second, a considerable block of copies are stored at the R&D center which produced the report, for use by people at the center. Then a considerable number of copies are sent to the National Technical Information Service for sale to the public. Also, a considerable number of these reports are stored at the central repository of the parent R&D agency, and a considerable number of copies are distributed to and stored in a group of regional repositories.

Report Storage - R&D Centers. The supply of a particular R&D report stored at the R&D center is usually kept in that center's technical library. This library often has branches in the major departments of the R&D center. For example, at LLL there is the main library and specialized branches of it in the Electronics Engineering Department, in the Mechanical Engineering Department, and in the Biomedical Department. The guide for the outsider should recommend that he ask for copies of the report from the main library, where they are usually kept. Some R&D centers also have a "documents" department where copies of the contractual reports are kept. This department may be a part of the center's library system. In that case also, the outsider should be advised that, in the R&D center's maze of technical information handling procedures, it is still the R&D center's library where he goes to get his hands on a copy of the contractual report.

The crucial consideration in getting at the reports stored in the R&D centers where they were produced is the problem, for the outsider, of getting access to the R&D center's technical library. In many cases the outsiders are arbitrarily barred from the center's library by its bureaucracy. To make regulations covering use of the R&D center's library by outsiders, and to take care of outsiders' requests to use it would undoubtedly increase the work of the library's staff. So the easy way out is to bar outsiders from access to the R&D center's library.

Fortunately, some agencies realize that this is an abuse of tax-supported facilities, so are opening the technical libraries of their R&D centers to the public. For example, NASA, as one aspect of compliance with the Freedom of Information Act, has issued regulations in the Federal Register governing the public's use of the technical libraries in its R&D centers (Federal Register, Part 1206 - Availability of Agency Records to Members of the Public). This will undoubtedly be followed in the future by other major R&D agencies.

Report Storage - Central Repositories. Each of the major R&D agencies has a number of centers in which its R&D programs are carried out. The Defense Department, Energy Department, and NASA each have a number of such R&D centers. Each of these parent agencies has a central repository where the contractual reports from all its R&D centers are kept. For the Defense Department it is the Defense Documentation Center at Alexandria, Virginia, near Washington, D.C.; for the Energy Department (formerly the AEC), the central repository for all its R&D reports is the Technical Information Center at Oak Ridge, Tennessee. NASA keeps a complete collection of its R&D reports at the Scientific and Technical Information Office, National Aeronautics and Space Administration, in Washington, D.C. Another important central repository is that of the National Technical Information Service in Springfield, Virginia, just outside Washington, D.C. A directory which lists all these central repositories and gives details of their report collections, facilities, etc. is provided in A DIRECTORY OF INFORMATION RESOURCES IN THE UNITED STATES FEDERAL GOVERNMENT (10).

These central repositories all brag about having millions of R&D contractual reports stored in their files, and about having their report collections computerized, so that one can find a desired report very quickly by computer search. They are huge, costly facilities built by the major R&D agencies to facilitate the research work of their R&D contractors. For that purpose they are good. But for technology transfer they are of little help, for this reason.

In effecting technology transfer it is crucial to get the R&D report, on a particular development that is to be transferred, within a year after that report was issued. After that it becomes difficult to find the people in the R&D center who made the development, and thus to obtain the technical information needed to convert that technical development into a commercial product quickly, easily and at low cost. This occurs because soon after a development is completed the team that developed it is often disbanded and its members transfer to other areas of work, retire, or leave the laboratory.

But the paper-pushing bureaucracy in Government R&D agencies is so cumbersome that it usually takes about a year, often longer, for a contractual report issued by an R&D center to be formally added to the parent R&D agency's central repository collection. Hence, the capability to retrieve a particular report from a report collection does not increase, no matter how huge and how computerized you make that collection, if the desired report has not yet been entered into the collection.

Report Storage - Regional Repositories. The U. S. Congress is concerned that Government documents be widely available to its citizens, at places where the citizens can conveniently get to them. It has therefore mandated that the Superintendent of Documents, through the Government Printing Office, establish regional repositories all over the country where citizens can find and study the Government-issued documents locally, conveniently, and at no cost.

The contractual reports issued by Government R&D agencies have been deemed Government documents that are included in this regional repository program. Each of the major R&D agencies has therefore worked out, with the Government Printing Office, a program that determines where that agency's contractual R&D reports will be stored in regional repositories throughout the country. The National Bureau of Standards has such a program; so do DoE, NASA, etc.

Mostly, these repositories are in selected public and university libraries (Fig. 8-5). For the outsider desiring to do technology transfer these regional repositories put a huge collection of technical material from Government R&D programs where he can reach them conveniently, can consult them without cost, and can study them at leisure and in depth.

This is the critical condition that makes the birth of good ideas for new products possible--when you can study the technical raw material out of which they are created at leisure and in depth. For example, Ernest O. Lawrence conceived his idea for the cyclotron when he was browsing through a scientific publication, one which was written in German, which language he did not understand; but he saw in that paper a diagram which contained the germ of the cyclotron concept.* The guide for the outsider through the maze

*See Ch. VII, Ref. (2).

DEPOSITORY LIBRARIES (Alphabetical by State and City)		CURRENT NBS PUBLICATIONS (SEE P. II) — PERIODICALS — JOURNAL OF RESEARCH (2-SECTIONS)			NONPERIODICALS										SUPERSEDI NBS (SEE PAGE II FOR MAY BE AVAILABL					
		A. PHYSICS AND CHEMISTRY	B. MATHEMATICAL SCIENCES	TECHNICAL NEWS BULLETIN	NATIONAL STANDARD REFERENCE DATA SERIES	BUILDING SCIENCE SERIES	HANDBOOKS	MONOGRAPHS	MISCELLANEOUS PUBLICATIONS/ SPECIAL PUBLICATIONS	APPLIED MATHEMATICS SERIES	PRODUCT STANDARDS (SEE COMM. STDS. & SIMP. PRAC. RECOMMEND.)	FEDERAL INFORMATION PROCESSING STANDARDS SERIES	CONSUMER INFORMATION SERIES	TECHNICAL NOTES	ANNUAL REPORT ONLY	JOURNAL OF RESEARCH (MONTHLY)	C. JOURNAL OF RESEARCH ENGINEERING AND INSTRUMENTATION	D. JOURNAL OF RESEARCH RADIO SCIENCE	CRPL IONOSPHERIC PREDICTIONS	BUILDING M STRUCTU	
	CALIFORNIA—Continued																				
	Occidental College, Mary Norton Clapp Library (1941)	✓						✓		✓				✓	✓	✓	✓			✓	
	Pepperdine University Library (1963)	✓	✓	✓			✓								✓		✓				
	University of California at Los Angeles Library (1932)	✓	✓	✓	✓	✓	✓	✓	✓	✓	✓			✓	✓	✓	✓	✓	✓	✓	
	University of California at Los Angeles, Law Library (1958)																				
	University of Southern California Library (1933)	✓	✓	✓	✓	✓	✓	✓	✓	✓	✓			✓	✓	✓	✓	✓	✓	✓	✓
Menlo Park	Department of the Interior, Geological Survey Library (1962)	✓	✓	✓	✓		✓	✓	✓	✓							✓	✓			
Montebello	Montebello Library (1966)						✓														
Monterey	Naval Postgraduate School Library (1963)	✓	✓	✓	✓		✓	✓	✓	✓				✓	✓		✓	✓	✓		
Monterey Park	Bruggemeyer Memorial Library (1964)														✓					✓	
Northridge	San Fernando Valley State College Library (1958)	✓	✓	✓	✓		✓	✓	✓	✓					✓		✓	✓			
Oakland	Mills College Library (1966)																				
	Oakland Public Library (1923)	✓	✓	✓	✓	✓	✓	✓	✓	✓	✓			✓	✓	✓	✓	✓	✓	✓	✓
Pasadena	California Institute of Technology, Millikan Memorial Library (1933)	✓	✓	✓	✓	✓	✓	✓	✓	✓	✓			✓	✓	✓	✓	✓	✓	✓	✓
	Pasadena Public Library (1963)	✓	✓	✓		✓	✓	✓	✓	✓	✓			✓	✓		✓	✓	✓		
Pleasant Hill	Contra Costa County Library (1964)						✓			✓	✓										
Redding	Shasta County Library (1956)	✓	✓				✓										✓	✓			
Redlands	University of Redlands, Armacost Library (1933)	✓	✓		✓		✓	✓	✓	✓				✓	✓	✓	✓	✓			✓
Redwood City	Redwood City Public Library (1966)					✓			✓	✓					✓						
Reseda	West Valley Regional Branch Library (1966)														✓						
Richmond	Richmond Public Library (1943)		✓					✓			✓				✓	✓	✓	✓			✓
Riverside	Riverside Public Library (1947)	✓	✓						✓							✓	✓	✓		✓	✓
	University of California at Riverside Library (1963)		✓		✓	✓		✓	✓	✓	✓			✓	✓		✓	✓			
Sacramento	California State Library (1895)—REGIONAL	✓	✓	✓	✓	✓	✓	✓	✓	✓	✓	✓	✓	✓	✓	✓	✓	✓	✓	✓	✓
	Sacramento City-County Library (1880)						✓	✓			✓				✓					✓	✓
	Sacramento County Law Library (1963)																				
	Sacramento State College Library (1963)	✓	✓		✓			✓	✓	✓					✓		✓	✓			
San Bernardino	San Bernardino County Free Library (1964)																			✓	
San Diego	San Diego County Library (1966)	✓	✓					✓									✓			✓	
	San Diego Public Library (1895)	✓	✓	✓	✓	✓	✓	✓	✓	✓	✓			✓	✓	✓	✓	✓	✓	✓	✓
	San Diego State College, Love Library (1962)	✓	✓	✓		✓		✓	✓	✓	✓			✓	✓		✓	✓			
	University of San Diego Law Library (1967)																				
San Francisco	Mechanics' Institute Library (1889)	✓	✓			✓	✓	✓	✓	✓	✓			✓		✓	✓			✓	✓
	San Francisco Public Library (1889)			✓	✓	✓			✓	✓	✓			✓	✓	✓			✓	✓	✓

Fig. 8-5. Partial list of repositories for technical information output of National Bureau of Standards.

of an R&D center's technical information handling practices should therefore include a good treatment of the regional repositories, located around that center, where stored collections of its R&D reports can be found.

Information Storage - Internal Technical Material

Internal Technical Information Material Storage Maze. Internal reports, drawings, specifications and manuals - these make up the major part of the technical information material produced at an R&D center, and they provide the most valuable material for effecting technology transfer. But each of these materials is produced by a separate group that has its own practices for preparing and its own places for storing its technical information material output. The storage of these materials is therefore a truly chaotic and confusing maze to the outsider looking to find them for use in the technology transfer process.

Internal Information Material Storage Maze - Case History. A particularly good example of how confusing and chaotic this maze can become is afforded by this technological development produced at LLL. The Radiochemistry Department at LLL had an R&D program for studying radioactive isotopes of very short half-lives--in the order of seconds. That is, their radioactivity dies out in a few seconds. So, to produce these short-life radioactive materials and do chemical tests on them to determine their chemical characteristics, the Radiochemistry Department had a system built in which a small sample of the chemical to be studied is inserted in a nuclear reactor and is irradiated. The sample is contained in a small cylindrical sample container which is carried, by a pneumatic-tube delivery system, from the point where it is loaded with the test sample, then is moved into the nuclear reactor and irradiated, and then it is moved at high speed (about 60 miles/hr) through the pneumatic sample-delivery system to an automatic chemical analysis station which performs the chemical tests required before the sample's radioactivity dies out. This entire system is completely automatic and the automatic control is provided by a microcomputer.*

To produce this system, the pneumatic sample-delivery system was designed and built by members of the Mechanical Engineering Department assigned to the Radiochemistry Department. They wrote the specifications, made the drawings, and wrote any internal reports on solving problems created by the system. These technical information materials are therefore stored in the ME Department.

---

*For details of this system see (11).

The mechanism for inserting the sample in the nuclear reactor, getting it properly irradiated, and then sent off to the automatic chemical analysis station, was designed and constructed by another group of mechanical engineers working in the Reactor Department of LLL.

The automatic chemical testing apparatus for testing the sample, after its irradiation, was planned and designed by the Radiochemistry Department, but was built for them by another group of mechanical engineers. They prepared the drawings and specifications on it, and these technical information materials are kept in the Mechanical Engineering Department.

Finally, the entire system's microcomputer-based automatic control system (11) was specified, drawn up, and built by a group of electronics engineers assigned to the Radiochemistry Department. They wrote the specs and made the drawings on it, so these materials are stored in the EE Department.

Thus the outsider wanting to get all the technical information available on this system, so he could transfer it economically to commercial application, would need a very good guide through this maze that would enable him to find where all the technical information produced on this system was stored and where copies of it could be obtained.

Internal Information Material Storage Guide. Fortunately, it is possible to provide a guide that would make it possible to create some measure of order in all this chaos and confusion. That would be the following procedure. Find the latest contractual report on the R&D program of which the technological development to be transferred is a part. For example, the annual report on the Short-Lived Radioactive Isotopes R&D program. These contractual reports have separate sections on the major technology developments included in the R&D program. These sections give not only the technical information on the development discussed, they also list the personnel involved in creating it; at least they give the name of the person in charge of the work described in that section of the contractual R&D report. This is the person well informed on the whole technological development discussed in that section of the contractual report, and he could be counted on to know who produced all the internal technical information material that was generated on it, as well as where their output of this material is stored. In short, to provide a guide to where the internal technical information on a particular technological development is stored, one should obtain the contractual report on the whole R&D program in which that technological development to be transferred is included. From the section in this report on that particular development one can find the names of the people who did the development work on it, hence who would know where all the internally generated technical material on that development--reports, drawings, specifications and manuals--could be found and studied.

For example, a valuable technological development for industry, produced in the R&D program at LLL on geothermal energy sources, is a process for treating the hot brine, that emerges from the geothermal wells, to prevent the chemicals contained in the brine from forming scale in the piping of the geothermal power generating system. This technological development is discussed in detail in a section, in the LLL GEOTHERMAL ENERGY PROGRAM STATUS REPORT,* Jan. 1976 - Jan. 1977, on "Scale Studies." It contains subsections on "Scale Composition and Microstructure," "Scaling Rates," and "Scale Control by Brine Modification." The engineers who conducted this work are all listed at the end of this section. Contact with them would enable the outsider to find out what internal reports, test reports, drawings, specifications, etc. were produced on this technological development at LLL and where in LLL to find them.

Information Dissemination - Contractual Material

The activity of disseminating the technical information gained in Government R&D programs to private industry is conducted, throughout all Government agencies, on the basis of the following theory.

All Government R&D activities have to be translated into R&D reports issued by the R&D center or agency that did the work. Each R&D report, as soon as it is issued, is sent to the major R&D agency's central repository, such as the DoD's Defense Documentation Center, NASA's Scientific and Technical Information Office, or the Department of Energy's Technical Information Center at Oak Ridge, Tennessee.

This central repository enters the R&D report into its computerized file of reports. The central repository's staff can then search for and retrieve that report, whenever called for, by entering the request into an appropriate computer search program.

Abstract Bulletins. At the same time, the central repository also enters that report's title and its abstract into an abstract bulletin of the R&D reports it received

---

*See Ch. VII, Ref. (4).

during the month. This monthly (or semi-monthly) abstract bulletin is published and sent out to a wide list of paid subscribers. Thus NASA publishes its monthly abstract service called STAR (Scientific and Technical Aerospace Reports); The Department of Energy's abstract service is called NUCLEAR SCIENCE ABSTRACTS. The Department of Defense sends practically its entire output of R&D reports to the National Technical Information Service (NTIS) for sale to the public, and NTIS publishes a bi-weekly abstract list of all the reports it received during that bi-weekly period, in a publication entitled THE GOVERNMENT REPORTS ANNOUNCEMENTS AND INDEX. The DoD therefore omits publishing an abstract list of its reports available to the public because that would be largely duplicating the list published by NTIS.*

The primary reason for this procedure is that publishing an abstract of each new R&D report in effect enables the entire public, nationwide, to learn of that report's title and contents, at the same time. Theoretically, this avoids giving anyone a time advantage in exploiting, commercially, the information contained in the R&D report.

This is the theory. The actual practice falls far short of this theoretical ideal for a number of reasons, as is made clear in the following discussion.

In the first place, the process of getting an R&D report entered into the central repository's computerized list takes a considerable length of time--in most cases about a year, or even longer. The result is that those private industry firms which carry out a Government R&D contract, and therefore write the R&D report on that project, have a time advantage of about a year in the use of the R&D report's contents. This can give them an enormous advantage in getting patents on the developments resulting from their R&D contract.

The other major flaw in this theory is that, in practice, a considerable portion of the unclassified R&D reports produced as a result of Government R&D programs are not put through this process. This process only covers those reports which the R&D agencies want sold to the public through NTIS. But the NTIS, though it receives the bulk

---

*These abstract publications include only unclassified reports. Announcements of classified reports are made by periodic report lists sent only to those authorized to receive classified publications.

of the unclassified R&D reports issued by Government agencies, gets these reports only from those Government agencies with which it has concluded agreements to serve as the sales agent for selling their reports to the public. There is no law which mandates that R&D agencies have to sell their R&D reports to the public through NTIS. So a considerable number of R&D agencies announce their reports to the public through the Government Printing Office's (GPO's) monthly list of documents it has published, and copies of those reports are sold to the public through GPO. For example, the National Bureau of Standards and the Bureau of Mines follow this procedure and get many of their reports announced and sold through GPO, not through NTIS.

Technology Transfer Centers. In addition to the dissemination of Government R&D results through the distribution of R&D reports, there is an extensive activity by Government R&D agencies in facilitating technology transfer directly, by bringing together the outsider looking for developments to transfer and the R&D centers which produce those developments.

A practically universal practice for doing this that has been developed in these R&D centers is to set up an "office" or "group" or "center" for technology transfer. This was first set up on a large scale by NASA, which established an office and an officer for technology transfer in each of its R&D centers. The Department of Defense then set up a "Consortium" of technology transfer sections in its R&D centers. Other R&D agencies have fallen in line with this and have added their "technology transfer offices" to this consortium. (Fig. 8-6.)

These offices of technology transfer have an organization called THE FEDERAL LABORATORY CONSORTIUM FOR TECHNOLOGY TRANSFER which is promoted by the National Science Foundation. Therefore, for a list of laboratories belonging to the Consortium, write to Federal Laboratory Consortium Manager, NATIONAL SCIENCE FOUNDATION - ISPT, Washington, D.C. 20550, Tel. (202) 634-7996. The activities of the Consortium are detailed in Ref. (12).

It is quite common for Government R&D centers to make provisions for the outsider to get the services of members of the center's staff as consultants, usually on a fee basis. It is very advisable for the outsider, seeking technology transfer information from an R&D center, to make his first contact with that center through its office of technology transfer and find out what these consultant arrangements are.

Laboratory	Location	Contact	Telephone
NASA George C. Marshall Space Flight Center	Huntsville, AL	Mr. Aubrey Smith	(205) 453-2224
Air Force Rocket Propulsion Laboratory	Edwards, CA	Mr. Gerald Sayles	(714) 553-2206 X32340
ERDA Lawrence Berkeley Laboratory	Berkeley, CA	Mr. Robert J. Morris	(415) 843-2740 X 5545
ERDA Lawrence Livermore Laboratory	Livermore, CA	Mr. R. Carroll Maninger	(415) 447-1100 X 7191
Forest Fire Laboratory	Riverside, CA	Mr. John R. Warren	(714) 787-1540
NASA Ames Research Center	Moffett Field, CA	Mr. Charles C. Kubokawa	(415) 965-5554
NASA Flight Research Center	Edwards, CA	Mr. Clinton T. Johnson	(805) 258-3311 X 568
NASA Jet Propulsion Laboratory	Pasadena, CA	Mr. John C. Drane	(213) 354-6420
Naval Civil Engineering Laboratory	Port Hueneme, CA	Mr. Eugene Early	(805) 982-4070
Naval Electronics Laboratory Center	San Diego, CA	Mr. Gerald E. Miller	(714) 225-7091
Naval Health Research Center	San Diego, CA	Mr. Milton Richlein	(714) 225-7393
Naval Pacific Missile Test Center	Pt. Mugu, CA	Dr. Samuel Brainin	(805) 982-8904
Naval Postgraduate School	Monterey, CA	Prof. J. W. Creighton	(805) 982-7051
Naval Undersea Center	San Diego, CA	Mr. Donald H. Courter	(714) 225-6251
Naval Weapons Center	China Lake, CA	Mr. George F. Linsteadt	(714)939-7325
Navy Personnel Research & Development Center	San Diego, CA	Mr. Allan A. Sjoholm	(714) 225-2712
U.S. Geological Survey (Western Region)	Menlo Park, CA	Mr. George E. Robinson	(415) 323-8111 X 2711
Dept. of Commerce Office of Telecommunications	Boulder, CO	Mr. Walter E. Johnson	(303) 499-1000 X 3501
Naval Underwater Systems Center	New London, CT	Mr. George Connolly, Jr.	(203) 442-0771 X 2908
Air Force Civil Engineering Center	Tyndal AFB, FL	Mr. Robert E. Brandon	(904) 283-6200
NASA Kennedy Space Center	Kennedy Space Ctr, FL	Mr. Raymond J. Cerrato	(305) 867-2780
Naval Coastal Systems Laboratory	Panama City, FL	Mr. William H. Williams	(904) 234-4204/05
Army Rock Island Arsenal	Rock Island, IL	Mr. Jack P. Manata	(309) 794-4518
Naval Weapons Support Center	Crane, IN	Mr. C. Dale Robinson	(812) 854-1282
Army Edgewood Arsenal	Aberdeen, MD	Mr. Frank Couch	(301) 671-3037
Army Harry Diamond Laboratory	Adelphi, MD	Mr. Clifford E. Lanham	(202) 394-2296
Army Human Engineering Laboratory	Aberdeen, MD	Mr. Donald Egner	(301) 278-4567
NASA Goddard Space Flight Center	Greenbelt, MD	Mr. Donald S. Friedman	(301) 982-6242
Naval Ship Research & Development Center	Bathesda, MD	Mr. David W. Nakonechny	(202) 227-1681
Air Force Cambridge Research Laboratory	Hanscom AFB, MA	Dr. John N. Howard	(617) 861-3161
Army Natick Development Center	Natick, MA	Dr. S. David Bailey	(617) 653-1000 X 2577
Dept. of Transp. – Transportation Systems Ctr.	Cambridge, MA	Mr. R. V. Giangrande	(617) 494-2486
Army Tank & Automotive Command Laboratory	Warren, MI	Mr. Ralph Trese	(313) 573-2319
NASA National Space Technology Laboratory	Bay St. Louis, MS	Mr. Roy S. Estess	(601) 688-3047
Army Cold Regions Research & Engineering Laboratory	Hanover, NH	Dr. Andre Assur	(603) 643-3200 X 237
Army Electronics Command	Fort Monmouth, NJ	Dr. Walter S. McAfee	(201) 535-1131
Army Picatinny Arsenal	Dover, NJ	Mr. Thomas C. Castorina	(201) 328-2560
National Aviation Facilities	Atlantic City, NJ	Mr. Joseph Rubino	(609) 641-8200 X 3373
Naval Air Engineering Center	Lakehurst, NJ	Mr. Charles O'Donnell	(201) 323-2648
Air Force Weapons Laboratory	Kirtland AFB, NM	Dr. Arthur Guenther	(505) 264-8561
ERDA Los Alamos Scientific Laboratory	Los Alamos, NM	Mr. Eugene Stark	(505) 667-6109
Air Force Rome Air Development Center	Griffiss AFB, NY	Mr. William S. Kelly	(315) 330-3046
Air Force Materials Laboratory	Wright-Patterson AFB, OH	MAJ Lawrence Fehrenbacher, USAF	(513) 255-3759
Air Force Wright Aeronautical Laboratory	Wright-Patterson AFB, OH	MAJ Robert Kosak, USAF	(513) 255-4052
NASA Lewis Research Center	Cleveland, OH	Mr. Paul Foster	(216) 433-4000
Naval Air Development Center	Warminster, PA	Mr. Jerome Bortman	(215) 672-9000 X 2140
EPA Environmental Research Laboratory	Narragansett, RI	Dr. Stanley H. Hegre	(401) 789-1071
NASA Johnson Space Flight Center	Houston, TX	Mr. John T. Wheeler	(713) 483-3809
Army Dugway Proving Grounds	Dugway, UT	Dr. Mortimer Rothenburg	(801) 522-2150 X 3314
Army Engineering Topographic Laboratory	Fort Belvoir, VA	Dr. Armando Mancini	(703) 664-2422
Army Mobility Equipment Research & Development Center	Fort Belvoir, VA	Dr. Karl Steinbach	(703) 664-5348
Army Night Vision Laboratory	Fort Belvoir, VA	Mr. Richard Fulton	(703) 664-3923
NASA Langley Research Center	Hampton, VA	Mr. John Samos	(804) 827-3281
NASA Wallops Flight Center	Wallops, VA	Mr. Gilmore H. Trafford	(804) 824-3411 X 2201
National Technical Information Service (NTIS)	Washington, DC	Dr. Sherman Gee	(202) 724-3374
Naval Oceanographic Office	Washington, DC	Mr. Clinton Beckner, Jr.	(202) 763-1115
Naval Research Laboratory	Washington, DC	Mr. Emanual Brancato	(202) 767-3046

T²

Federal Laboratory Consortium for Technology Transfer

Fig. 8-6. Early Membership List, Federal Laboratory Consortium For Technology Transfer.

Direct Dissemination. A further interesting and valuable aspect of dissemination that should be called to the outsider's attention, in the guide on technical information practices prepared for him, is that some R&D agencies have report disseminating projects that are comparable to, and as valuable as, the National Technical Information Service. For example, the National Bureau of Standards conducts an extensive operation for informing the public about R&D reports it issues, and it carries on a sizable operation for selling or giving away copies of these reports to the public. Both these activities are described in Ch. IX, Case History - Guide Through the National Bureau of Standards.

Summary

To find, in an R&D center, the technical information needed to effect the transfer of one of its technological developments quickly, easily and at low cost, one must first realize that there are two major categories of technical information material produced in those R&D centers. One is the category of contractual R&D reports; the other is the category of internal technical information material.

Summary - Contractual Reports

Generation. On all R&D projects funded by the Federal Government, the final step required is the submission of an R&D report which tells what was accomplished and and how it was achieved.

Since many R&D projects run on for years, these reports are often submitted as annual progress reports. And because they are used as the basis for evaluating the performance on projects funded in the past and for funding projects in the future, these reports are often professionally done, produced by professional writer/editors working in close cooperation with the top scientists and engineers involved in the R&D project. On the whole, the generation of contractual reports is organized, systematic, with abundant resources provided to accomplish their production. Moreover, the generation of contractual reports is done quite uniformly throughout the R&D centers of many R&D agencies.

Storage. The storage of contractual reports is also well organized and systematic. Each R&D center stores a supply of its contractual reports in the center's technical library. These contractual reports are also stored in the principal repository each agency has for storing the contractual reports from all its R&D centers. Each major R&D agency also has regional repositories where its collection of contractual reports can be consulted locally.

Dissemination. Dissemination of contractual reports is also well organized, systematic, and done uniformly on a large scale throughout the Government R&D agencies. This is accomplished by two major measures:

1. First, abstract bulletins are published monthly (or semi-monthly) by each of the major R&D agency central repositories. These bulletins present abstracts of the contractual reports added to their respective collections during that month. This notifies the public that the reports listed are now available to them and indicates their content by the abstract.

2. The reports listed in the abstract bulletins are sold to the public through the National Technical Information Service.

Summary - Internal Technical Information Material

Generation: Internal Information Material. The technical reports issued by Government R&D centers or agencies, and which the public generally considers as the main source of technical information useful in effecting technology transfer, constitute only a fragment of the technical information generated in Government R&D centers. The total output of technical information in those centers comprises contractual reports, internal reports, drawings, specifications, and equipment manuals. All these different types of documents contain technical information that is needed for effecting transfer of a technological development, and the information in all of them must be gathered and integrated to effect a technology transfer economically--without reinventing the wheel. The guide through the technical information handling maze of the R&D center must therefore make clear how all these different types of technical information sources are generated, stored and disseminated.

Internal Reports. Of far greater number than the contractual reports are the internal reports generated on each technological development created at the R&D center. The internal reports are often engineering studies, to find the solutions to problems that crop up in the course of producing a particular technological development. They often contain one valuable thing. That is, they often show the process of reasoning for particular steps taken and this can be the most helpful thing of all to the outsider striving to convert a particular technological development from a Government R&D program into a successful commercial product. The guide through the maze of technological information handling at an R&D center should therefore give a clear picture of where and how the outsider can find out which of these internal reports have been produced, and how he can get access to their contents.

Drawings. The drawings produced to achieve a particular technological development of course contain details of manufacture and construction that can save the person, who is planning to effect a transfer of that development, a great deal of time and money. A complete set of drawings is usually produced in the drafting section of the engineering department working on that development. In different engineering departments the drawings are produced and numbered according to different procedures. The guide through the maze of an R&D center's information handling procedures should therefore treat these in detail for each department.

Equipment Manuals. Included in the drawing packages produced at LLL is a document called the "Description and Instruction" document (D&I) which gives a description of how the equipment unit works and instructions for its installation, operation, maintenance, and servicing.

These D&I's are equivalent to the equipment manual usually produced by the armed services for each piece of equipment the soldier or sailor uses. They are thus immensely valuable for technology transfer because, if you know how the equipment operates and how it should be serviced, you have the bulk of the technical information needed to effect its conversion to a commercial product.

Specifications. In practically all R&D centers a specification is drawn up on a technological development before it is built. These specifications give the performance requirements that the system, subsystem, or equipment unit that is to be built shall meet. But a good specification contains, besides performance requirements, certain utility requirements. These are the requirements that the equipment to be built shall meet to make its operation safe, convenient, etc. Obtaining the specification on the equipment unit to be transferred can therefore be very helpful in converting it to a successful commercial product.

Storage: Internal Technical Information Material. Internal reports, drawings, specifications and manuals - these make up the major part of the technical information material produced at an R&D center, and they provide the most valuable material for effecting technology transfer. But each of these materials is produced by a separate group that has its own practices for preparing and its own places for storing its technical information material output. The storage of these materials is therefore a truly chaotic and confusing maze to the outsider looking to find them for use in the technology transfer process.

Fortunately, it is possible to provide a guide that would make it possible to create some measure of order in all this chaos and confusion. That would be the following procedure. Find the latest contractual report on the R&D program of which the technological development to be transferred is a part. For example, the annual report on the Short-Lived Radioactive Isotopes R&D program. These contractual reports have separate sections on the major technology developments included in the R&D program. These sections give not only the technical information on the development discussed, they also list the personnel involved in creating it; at least they give the name of the person in charge of the work described in that section of the contractual R&D report. This is the person well informed on the whole technological development discussed in that section of the contractual report, and he could be counted on to know who produced all the internal technical information material that was generated on it, as well as where their output of this material is stored.

Dissemination: Internal Information Material. As noted above, the most valuable material for effecting a technology transfer quickly and at low cost is the wealth of internal information material that is generated at an R&D center in the course of creating the technological development to be transferred--the internal reports, the drawings, specifications and manuals. But the generation and storage of this material at the center are usually fragmented, scattered, and uncoordinated. There are three channels that will enable the outsider to gain access to, gather up, and tie together this material.

1. Dissemination: Offices of Technology Transfer. Practically all R&D centers now have "offices of technology transfer" that are intended to help the outsider to find and obtain this internal technical information material. Where this "office of technology transfer" is adequately staffed with qualified and conscientious personnel,

they will give the outsider substantial help in finding his way through the maze of the R&D center's internal technical information handling processes. However, as is often the case in Government operations, they are overloaded and understaffed. In that case they do not dig up comprehensive answers to his needs for internal reports, drawings, specifications, and manuals. All the outsider gets is a brush-off through a routine mailing of one or more pieces of paper they have already available on the shelves. As demand for technology transfer information grows and complaints about incompetent service increase, this situation will hopefully be improved by better staffing.

2. <u>Dissemination: R&D Center Libraries</u>. Another avenue of access to the rich store of internal technical information available at an R&D center is the opening up of the center's technical library to the public. Here the material the transferee needs--reports, drawings, specifications and manuals--are often stored, or the librarian can direct him to the people who know where it can be found.

   NASA has already opened the libraries at its R&D centers to the public. Other agencies are reluctant to do so because opening a library to the outsider increases the workload on the library's staff. As demand for technology transfer information increases the R&D bureaucracies will inevitably have to expand this avenue of access for the outsider to the technical information material he needs.

3. <u>Dissemination: Guides to Technical Information Handling</u>. The third avenue of access for the outsider to get and use the internal technical information material generated and stored in an R&D center is to prepare a guide through the maze of the center's technical information handling activities. The center's staff of technical writers and editors are qualified to prepare such a guide. This chapter furnishes them the information needed to do it--a comprehensive picture of how reports, drawings, specifications and manuals are generated and stored in an R&D center. The R&D centers are therefore in good position to prepare such guides. It is to be

hoped that, in the future, they will pave the way for greater technology transfer activity by producing them.

REFERENCES

1. Use RASTER OSCILLOSCOPES for Faster Time Measurements, Richard P. Rufer and William A. Karlotski, ELECTRONICS, Dec. 28, 1962, pp. 38-42.

2. Description and Instructions, Raster Oscilloscope, Dual, L-10 Model 2, Document LE8188-2M2, Electronics Engineering Department, LLL, P. O. Box 808, Livermore, CA 94550, 32 pages.

3. L-300 RASTER OSCILLOSCOPE OPERATION AND MAINTENANCE MANUAL. J. C. Lawson, Document M-080, LEA 75-003-002, Feb. 2, 1977, 116 pp. Electronics Engineering Dept. LLL, P. O. Box 808, Livermore, CA 94550.

4. Specification, ELECTRON MICROSCOPE, EE Dept., LLL LES 11403, 10 pp. Issued by Standards and Specifications Section, Electronics Engineering Dept., Lawrence Livermore Laboratory, P. O. Box 808, Livermore, CA 94550.

5. A Systematic Procedure for Preparing Specifications on Electronic Instrumentation and Control Systems, Hyman Olken. IRE TRANSACTIONS ON ENGINEERING MANAGEMENT, Sept. 1961, pp. 143-153.

6. Procurement Specifications for High Technology Research Equipment, Jack W. Spencer. PROCEEDINGS, 26th International Technical Communications Conference, Society for Technical Communications, Los Angeles, CA, May 16-19, 1979, pp. W-149 to W-153.

7. Electronics Engineering Standards Manual, 3000 pp. Issued by Electronics Engineering Dept., Lawrence Livermore Laboratory, P. O. Box 808, Livermore, CA 94550.

8. Microcomputers - From the User's Viewpoint, Hyman Olken. RESEARCH AND DEVELOPMENT, Oct. 1974, pp. 16-24.

9. A Stores Stock Microcomputer System: The MCS-8 8-Bit Processor Controller. Electronics Engineering Dept. Report LER 73-102301, Nov. 7, 1973. E. Fisher, J. Spann, H. Olken, R. Goodman, S. Nielsen. Issued

by Manufacturing and Materials Engineering Group, Electronics Engineering Dept., Lawrence Livermore Laboratory, P. O. Box 808, Livermore, CA 94550.

10. A DIRECTORY OF INFORMATION RESOURCES IN THE UNITED STATES FEDERAL GOVERNMENT. Revised Edition, 1974. Science and Technology Division, National Referral Center, Library of Congress, Washington, D.C. Order from Supt. of Documents. U.S. Government Printing Office, Washington, D.C. 20402. Stock No. 3000-00067, Price $4.25.

11. Fast Rabbit/Rapid Chemistry - A Control System For Automated Chemical Analysis, S. Bourret. ELECTRONIC ENGINEERING DEPARTMENT QUARTERLY REPORT No. 3, 1975, UCRL 50025-75-3, July 24, 1975, pp. 6-9. Available from NTIS, Springfield, VA 22151.

12. The Federal Laboratory Consortium For Technology Transfer. Robert C. Crawford. DEFENSE SYSTEMS MANAGEMENT REVIEW. Winter, 1979. Vol. 2, No. 1, pp. 53-64. Available from National Technical Information Service (NTIS), Springfield, VA 22151.

Ch. IX

CASE HISTORY - GUIDE THROUGH THE NATIONAL BUREAU OF STANDARDS

Outline

	Page
The Composite Guide	95
Survey Article	95
Annual Report	97
NBS Organization	97
NBS Technical Activities	98
Formal Guide	99
Guide Through the NBS Technical Information Program	100
NBS Publications Newsletter	101
Other Aspects of NBS Technical Information Handling	101
Summary	103

Appendices

App. 9-1 Computer Controlled Satellite Clock 126

Ch. IX

CASE HISTORY - GUIDE THROUGH THE NATIONAL BUREAU OF STANDARDS

The Composite Guide

A guide that will enable the outsider to find his way through the maze of a Government R&D center, so he can obtain the technical information needed to effect a technology transfer, need not consist of a single document. It can be composed of a group of documents or procedures that contribute, singly and in combination, to the building of such a guide. In the case of the National Bureau of Standards this group consisted of:

1. A survey article
2. An annual report
3. A formal guide
4. A guide through its technical information program.

The Survey Article

A good way to begin the guide through the maze of an R&D center is to start with a comprehensive article that enables the outsider to see the R&D center as a whole. That is, one that gives an indication of the R&D center's history, its mission, its organizational framework, its areas of scientific and technological activity, and its scientific and technical competences and accomplishments.

Such an article was provided for NBS in the article entitled The National Bureau of Standards - America's Inquisitive Agency, published in the September-October 1978 issue of the SMITHSONIAN Magazine. This is a posh, 4-color, 18-page PR triumph. It comprises two parts, one of which covers the Bureau's original and basic activity--insuring the accuracy of measures used in commerce and industry. The second part covers the Bureau's branching outward into modern fields of science and technology.

Although it emphasizes the publicity-catching aspects of the Bureau and its work, this article gives a good intellectual look at the Bureau and makes one very significant point that distinguishes NBS from other national laboratories. That is, that the Bureau itself pretty much determines its own boundaries for its scientific activities. After pointing out the Bureau's basic mission, as defined by Congress, to establish and refine the basic standards of weights and measures, the article goes on to point out:

> ". . . But one extra clause in its charter both handed the Bureau another responsibility and, as it turned out, bestowed on it enormous discretion to inquire and to act."

This paragraph (the "clause" referred to above) enjoins NBS to determine the

> "physical constants and the properties of materials when such data are of great importance to scientific or manufacturing interests and are not to be obtained of sufficient accuracy elsewhere."

This is a blank check given by Congress to the Bureau to define its own limits for its activity. As the author puts it:

> ". . . it (the Bureau) has come to take that extra clause as a green light for research into almost any problem that appears to need solution."

In pursuit of this self-bounding charter, the NBS has moved into modern high technology fields such as computers, robots, and acoustic devices for imaging the insides of the human body. This article portrays those areas of NBS activity as well as its more mundane and basic activity of setting up and improving the standards and techniques for measurement in all areas of science and industry.

Such an article provides an introduction to an R&D center that both interests and informs the outsider. It is commonly used by R&D centers as the opening wedge to the guide through the R&D center's maze, only other agencies do not do it on so elaborate a scale. Often such an introduction is provided through the medium of statements defining the R&D center and its operations at hearings before Congressional committees. NASA particularly, and formerly the AEC, followed this route (1)(2)(3).

Other Government agencies, such as the Army, provide introductory statements about their R&D centers by articles published in their own agency publications. Thus the Army has for years published such articles on many of its R&D centers (e.g., smart weapons development center, at Huntsville, Alabama) in the Army R&D Magazine (4).

Such a comprehensive survey article is a vital first step in providing a guide for the outsider through the maze of an R&D center. This article on NBS presents a good example of what such an article should cover and how it should cover it.

Annual Report

The annual report of an R&D center is the major part of the guide, through the maze which that center presents to an outsider, because it gives a detailed account of the R&D activities at the center and of the center's organizational structure. A good example is the Annual Report of the National Bureau of Standards for the fiscal year 1978.*

NBS Organization. This 85-page report is particularly valuable as a guide to the outsider because it points out that NBS is radically different from other R&D centers, and how this determines its R&D activities and output. Thus the NBS is not devoted solely to R&D work as most R&D centers are, because NBS is assigned, by Congress, a number of different missions. The major one is to develop standards and methods in all areas of measurement--weight, length, temperature, etc. In this area it is an R&D center. But Congress has also assigned to NBS the job of being a technical service agency to other Government agencies and to do their technical chores, like standardizing their computers and their data processing procedures or helping the FBI to automate its fingerprint identification system. The report therefore has a three-page listing of the acts of Congress which assign NBS the things it is to do.

Another major difference between NBS and other R&D centers or agencies, made clear by this annual report, is that NBS has a large responsibility to see that its standards and other developments, in fields like weights and measures, are transferred to and adopted by local Government bodies. This is covered by a five-page section on Science and Technology Transfer: Services and Special Programs. This section includes such things as a Calibration Service, to help other laboratories insure the accuracy of their instruments, and the Standard Reference Materials Service, which provides science and industry with materials of specific properties required in measurement applications.

Another useful feature of this annual report to the outsider is a three-page directory which lists the various divisions of NBS, thus indicating its organizational structure. The directory gives the phone numbers as well as

*Ch. VI, Ref. (2).

organizational location of different individuals, thus making it easy for outsiders to contact the people from whom technology transfer information has to be obtained.

NBS Technical Activities. As noted above, under the Survey Article (p. 95) Congress has given NBS practically a blank check to determine its own boundaries for its scientific activities. A reading of this report gives the distinct impression that, as a result, NBS has spread itself into many areas of science and technology, and has consequently attained "also-ran" status in many and leadership status in only one--the measurement of time and frequency. In this one field it has also attained a high-technology spin-off that is valuable to both science and industry, and which is described in detail in Appendix 9-1..

However, it is leadership in the foremost frontiers of science that nowadays gives industrial countries the foundation on which to build high technology industries. Lack of such leadership at NBS could therefore be a contributing factor to America's recently falling behind other industrial nations in the competition for high technology industry.

Apparently NBS has become aware of this and is planning to do something about it, as indicated by the following statements in this annual report.

> "We have now launched a special effort to renew and enhance our technical competences in research areas critical to the nation and basic to our goals. This strengthening of technical excellence is designed to ensure that the Bureau can continue to provide the underpinning essential to our ability as a nation to make studies in top priority and emerging research fields." (From Foreword, by Director of NBS, on p. 3 of this annual report.)

Specifically, this leads to the following activities in the current and future years.

> "The Bureau's ability to fulfill its multiple missions and provide the measurement foundations for technological innovation is heavily dependent on the continued maintenance and enhancement of its technical competence. Thus, NBS last year embarked on a program to strengthen key research. . . . Research fields selected for this multi-year program will involve long-term theoretical and experimental studies upon which NBS and its clients can rely for expertise. Special competence building initiatives are being undertaken where scientific solutions compatible with the NBS mission appear possible

> and where the underlying scientific foundation appears ripe for advancement." (1978 NBS Annual Report, p. 7.)

One such area that is "ripe for advancement" has already been selected and work on it started. It is considered of such importance that it is the first item treated in this report's detailed discussion of the Bureau's R&D programs. As stated on page eight of this report:

> "Surface science, one of the fields selected for the first installment of the special competence enhancement program, illustrates the fundamental importance and nature of the research areas being given this special attention. . . . Our ability to understand and control processes in such diverse areas as catalysis, adhesion, corrosion, lubrication, frictional wear, bonding, and oxidation and our use of an array of technologies, including microelectronics and many energy production processes, rests upon advances in surface science. . . ."

How the NBS scientists will proceed to turn this vision of future high technology into reality is indicated as this section of its annual report continues (p. 8).

> "The properties of surfaces are known to differ from the properties of the bulk of many materials. To a great extent, these surface properties are determined by the chemical identity and the position of individual atoms and groups of atoms at or near the surface. Scientists working in this field are seeking to understand surface properties and processes at the atomic level. One of their principal challenges is to develop improved measurement methods with the sensitivity, specificity, and accuracy needed to identify surface atoms and atom arrangements in both laboratory and practical situations."

In short, the annual report is the key factor in a guide for the outsider through the maze of an R&D center because it gives full information on the R&D programs and other technical activities at the center and on the center's organizational structure. This NBS annual report is a particularly good example, because the R&D programs at a center can change drastically and this report shows how such drastic changes may come about.

Formal Guide

The pinnacle of the composite guide for the outsider through the maze of an R&D center is a document specifically directed to the outsider, and concerned with giving him specific directions for finding the technology transfer information he needs in that R&D center. A good example is the 45-page booklet produced by NBS entitled SCIENCE ON ITS WAY TO WORK - Activities of the National Bureau of Standards.

This booklet describes the NBS activities from the user's point of view. To achieve this it treats the major areas of NBS activity so as to point out what help the outsider can get from them. This makes for a well planned organization of the booklet into these major divisions or chapters: Support to Industry; Support to Other Government Agencies; On the International Scene, a section on support of U.S. foreign policy through support of international technological activities; and a chapter on And in the Public Interest, which describes NBS effort toward combating pollution, enhancing radiation safety and consumer product safety. Building the Foundations is the title of the chapter that covers NBS activities in the scientific research field and the application of that work to industry. A chapter on Funds and Facilities points out which NBS facilities "are available for use by the Scientific and Engineering Communities" and how to get the use of them. A good background piece is the chapter on Summary of Legislation, which gives the outsider a good view of how NBS got involved in so many different areas of science and technology

There are few similar documents in the huge outpouring of material from the nation's many R&D centers, so this one can well serve as a model for those preparing the guide for the outsider through the maze of an R&D center.

Guide Through the NBS Technical Information Program

The groups who work in different scientific and technological areas at NBS have different audiences. Each group beams the output of its work to its special audience. The result is an outpouring of information from NBS in various forms. One part is in the form of conventional contractual reports which are sold to the public through NTIS or through the Government Printing Office. A great many NBS technical papers are presented at conferences, so are published in the proceedings of those conferences. Another major part of the technical information output of NBS is in the form of articles published in various technical and professional magazines and journals. Altogether, there is a huge outpouring of technical information from NBS that is funneled to the respective audiences through a variety of channels. This indeed makes for a maze that the outsider must find his way through to get the information he needs to effect a technology transfer.

NBS Publications Newsletter. Fortunately, the NBS is keenly aware of this, so provides a guide through its technical information program called the NBS PUBLICATIONS NEWSLETTER. It is a bimonthly publication which gives abstracts of all the different varieties of documents and publications issued by NBS or authored by its staff. This could be a chaotic hodge-podge that would confuse any would-be technology transferee, but instead it is a very useful guide to the outsider because of the following features.

1. The abstracts are classified by subject areas, so if you are interested in Measurement Science and Technology, you will find papers, articles, reports, conference proceedings, even book reviews on that area of technology all in one group. Other subject groupings are Electronics Technology, Lasers and their Applications, Computer Science and Technology, etc.

2. The abstracts give full information on the documents from which they are taken. If a scientific journal, full information is given on the publication, its issue, date, page numbers, etc. If a publication of NBS, it tells you whether you can order it from NTIS or from the Government Printing Office (Superintendent of Documents), the ordering number, the number of pages, and the price.

3. The abstracts stress technology transfer information. It is characteristic of NBS abstracts in the PUBLI CATIONS NEWSLETTER that they stress information that would be helpful in technology transfer. A good example is the abstract (No. SP 400-54) on page 7 of the August 1979 issue of the NEWSLETTER, on an area transistor tester which offers construction notes, circuit schematics and mechanical drawings to facilitate its construction.

4. Other NEWSLETTER features. A back-cover page (Fig. 9-1(B)) gives a directory of the different types of NBS publications, where they can be obtained, and at what prices. This is very helpful to the outsider in interpreting the information in each abstract as to where the document abstracted was published, how a copy can be obtained, at what price, etc.

A characteristic of primary value is that this NBS PUBLICATIONS NEWSLETTER can be obtained at no charge, merely by addressing a request to the editor, in the Technical Information and Publications Division, NBS, Washington, D.C. 20234.

Other aspects of NBS technical information handling. There are also these two valuable general features of the

U.S. DEPARTMENT OF COMMERCE
National Bureau of Standards

NBS Publications Newsletter

(A)

For Librarians, Documentalists, and Science Information Specialists.

JANUARY 1980 NO. 80-1

NBS ANNUAL PUBLICATIONS CATALOG

Publications of the National Bureau of Standards. 1978 Catalog. A Compilation of Abstracts, and Key Word and Author Indexes (SP 305 Supplement 10) is now available from the Superintendent of Documents, GPO. (See page 14.) This volume lists publications of the Bureau issued between January 1-December 31, 1978. Also included are some earlier papers omitted from the previous annual catalog, together with general information and instruction about NBS publications.

This catalog will be of special interest to readers of this newsletter because it is a permanent integrated source of information on all NBS publications issued in 1978. It also includes information on the availability of previous annual NBS publications catalogs, plus a broad overview of the entire NBS publications program. We especially call this item to your attention.

COMPUTER SCIENCE AND TECHNOLOGY

Three additional publications have been issued in the NBS Special Publication 500-subseries dedicated to computer science and technology. Computer Science and Technology: Measurement of Interactive Computing: Methodology and Application (SP 500-48 on page 2) addresses the measurement of interactive computing, focusing on the performance of the user and the system in individual interaction sessions. The methodology of this study is a major contribution, providing a quantitative way to investigate a variety of phenomena associated with interactive computing. Computer Science and Technology: Modeling and Measurement Techniques for Evaluation of Design Alternatives in the Implementation of Database Management Software (SP 500-49 on page 3) speaks to the problem posed by the fact that the traditional hardware/software developmental approach, which consists in actually building, and then modifying, prototype systems, is no longer practical because of the great expenses incurred. It is necessary to determine the gross feasibility of an approach prior to the commitment of resources. The research reported in this publication demonstrated that proposed designs can indeed be evaluated through the use of performance prediction models.

Computer Science and Technology: Computers, Personnel Administration and Citizen's

Comments, or requests to be placed on distribution for this newsletter, may be sent to: NBS Publications Newsletter, Editor: John Rochford, Technical Information and Publications Division, National Bureau of Standards, Washington, D.C. 20234

NBS TECHNICAL PUBLICATIONS (B)

PERIODICALS

JOURNAL OF RESEARCH—The Journal of Research of the National Bureau of Standards reports NBS research and development in those disciplines of the physical and engineering sciences in which the Bureau is active. These include physics, chemistry, engineering, mathematics, and computer sciences. Papers cover a broad range of subjects, with major emphasis on measurement methodology and the basic technology underlying standardization. Also included from time to time are survey articles on topics closely related to the Bureau's technical and scientific programs. As a special service to subscribers each issue contains complete citations to all recent Bureau publications in both NBS and non-NBS media. Issued six times a year. Annual subscription: domestic $17; foreign $21.25. Single copy, $3 domestic; $3.75 foreign.

NOTE: The Journal was formerly published in two sections: Section A "Physics and Chemistry" and Section B "Mathematical Sciences."

DIMENSIONS/NBS—This monthly magazine is published to inform scientists, engineers, business and industry leaders, teachers, students, and consumers of the latest advances in science and technology, with primary emphasis on work at NBS. The magazine highlights and reviews such issues as energy research, fire protection, building technology, metric conversion, pollution abatement, health and safety, and consumer product performance. In addition, it reports the results of Bureau programs in measurement standards and techniques, properties of matter and materials, engineering standards and services, instrumentation, and automatic data processing. Annual subscription: domestic $11; foreign $13.75.

NONPERIODICALS

Monographs—Major contributions to the technical literature on various subjects related to the Bureau's scientific and technical activities.

Handbooks—Recommended codes of engineering and industrial practice (including safety codes) developed in cooperation with interested industries, professional organizations, and regulatory bodies.

Special Publications—Include proceedings of conferences sponsored by NBS, NBS annual reports, and other special publications appropriate to this grouping such as wall charts, pocket cards, and bibliographies.

Applied Mathematics Series—Mathematical tables, manuals, and studies of special interest to physicists, engineers, chemists, biologists, mathematicians, computer programmers, and others engaged in scientific and technical work.

National Standard Reference Data Series—Provides quantitative data on the physical and chemical properties of materials, compiled from the world's literature and critically evaluated. Developed under a worldwide program coordinated by NBS under the authority of the National Standard Data Act (Public Law 90-396).

NOTE: The principal publication outlet for the foregoing data is the Journal of Physical and Chemical Reference Data (JPCRD) published quarterly for NBS by the American Chemical Society (ACS) and the American Institute of Physics (AIP). Subscriptions, reprints, and supplements available from ACS, 1155 Sixteenth St., NW, Washington, DC 20056.

Building Science Series—Disseminates technical information developed at the Bureau on building materials, components, systems, and whole structures. The series presents research results, test methods, and performance criteria related to the structural and environmental functions and the durability and safety characteristics of building elements and systems.

Technical Notes—Studies or reports which are complete in themselves but restrictive in their treatment of a subject. Analogous to monographs but not so comprehensive in scope or definitive in treatment of the subject area. Often serve as a vehicle for final reports of work performed at NBS under the sponsorship of other government agencies.

Voluntary Product Standards—Developed under procedures published by the Department of Commerce in Part 10, Title 15, of the Code of Federal Regulations. The standards establish nationally recognized requirements for products, and provide all concerned interests with a basis for common understanding of the characteristics of the products. NBS administers this program as a supplement to the activities of the private sector standardizing organizations.

Consumer Information Series—Practical information, based on NBS research and experience, covering areas of interest to the consumer. Easily understandable language and illustrations provide useful background knowledge for shopping in today's technological marketplace.

Order the **above** *NBS publications from: Superintendent of Documents, Government Printing Office, Washington, DC 20402.*

Order the **following** *NBS publications—FIPS and NBSIR's—from the National Technical Information Services, Springfield, VA 22161.*

Federal Information Processing Standards Publications (FIPS PUB)—Publications in this series collectively constitute the Federal Information Processing Standards Register. The Register serves as the official source of information in the Federal Government regarding standards issued by NBS pursuant to the Federal Property and Administrative Services Act of 1949 as amended, Public Law 89-306 (79 Stat. 1127), and as implemented by Executive Order 11717 (38 FR 12315, dated May 11, 1973) and Part 6 of Title 15 CFR (Code of Federal Regulations).

NBS Interagency Reports (NBSIR)—A special series of interim or final reports on work performed by NBS for outside sponsors (both government and non-government). In general, initial distribution is handled by the sponsor; public distribution is by the National Technical Information Services, Springfield, VA 22161, in paper copy or microfiche form.

BIBLIOGRAPHIC SUBSCRIPTION SERVICES

The following current-awareness and literature-survey bibliographies are issued periodically by the Bureau:

Cryogenic Data Center Current Awareness Service. A literature survey issued biweekly. Annual subscription: domestic $25; foreign $30.

Liquefied Natural Gas. A literature survey issued quarterly. Annual subscription: $20.

Superconducting Devices and Materials. A literature survey issued quarterly. Annual subscription: $30. Please send subscription orders and remittances for the preceding bibliographic services to the National Bureau of Standards, Cryogenic Data Center (736) Boulder, CO 80303.

Fig. 9-1. Typical issue, the NBS Publications Newsletter. (A) Front cover; (B) back cover.

technical information handling process at NBS that are reflected in this NEWSLETTER.

The NBS often acts as a coordinator to enhance the work of all R&D centers at work in the same field by bringing them together at a seminar, symposium or conference on that field. This obviously enriches the work of the participants by an exchange of know-how in that field. In addition, the NBS always publishes a Proceedings of all the reports presented at such a seminar or symposium. This gives the outsider interested in technology transfer a rich opportunity to take advantage of the work in not only one, but in a number of R&D centers active in that one area of technology.

The other characteristic of the NBS technical information handling process that helps guide the outsider to finding technology transfer information developed in NBS is that all items (other than the seminar proceedings) listed in the NEWSLETTER are authored by members of the NBS staff. Hence, if the outsider has difficulty in getting copies of the papers the abstracts of which appeared in the NEWSLETTER, he can refer to the Directory of NBS (given in the Annual Report or in the Formal Guide) and call up the author of the paper or the editor of the NEWSLETTER. In fact, the NEWSLETTER always has a note saying that "papers or articles published in non-NBS media can often be obtained directly from the author."

Thus, for example, the author of this book wanted to get the paper which is presented in Appendix 9-1. He wrote to the editor of the NEWSLETTER and was advised to contact the author of that paper in Boulder, Colorado laboratory of NBS. He did, and soon received a reprint of the paper.

In short, the NBS PUBLICATIONS NEWSLETTER not only gives you good abstracts but makes the papers you read about in the abstracts obtainable. This is of tremendous value in technology transfer.

Summary

Altogether, one notes, from all the guides provided by NBS to help the outsider find his way through the maze that the Bureau represents, that the NBS is technology transfer-minded and arranges its activities so as to help the outsider obtain the technical information needed to effect technology transfer. Furthermore, the component elements of the composite guide through the NBS maze are good models for the technical writer/editor to follow when preparing the guide through the maze of his own R&D center.

REFERENCES

(1) ANNUAL REPORT TO CONGRESS OF THE ATOMIC ENERGY COMMISSION for 1970. U.S. Atomic Energy Commission, Jan. 1971. 324 pp. Available from U.S. Government Printing Office.

(2) 22nd SEMIANNUAL REPORT TO CONGRESS July 1-Dec. 31, 1969. National Aeronautics and Space Administration, Washington, D.C. 224 pp. Available from U.S. Government Printing Office.

NOTE: Other reports to Congress in the early and mid-1960's have more information on research facilities of those agencies.

CONSULT ALSO:

(3) SPACE PROGRAM BENEFITS. Hearing Before the Committee on Aeronautical and Space Sciences, U.S. Senate, 91st Congress, Second Session, April 6, 1970, Washington, D.C. 361 pp. Available from U.S. Government Printing Office.

(4) Used to be ARMY RESEARCH and DEVELOPMENT MAGAZINE, now ARMY RESEARCH, DEVELOPMENT and ACQUISITION MAGAZINE. Published bi-monthly by the Development and Engineering Directorate (DRCDE), HQ U.S. Army Material Development and Readiness Command, Alexandria, VA.

APPENDICES

Page

Ch. III

App. 3-1. Uninterruptable Power System 106

App. 3-2. Remotely Controlled Seismic Network 108

App. 3-3. Transportable Test Instrumentation System 110

Ch. IV

App. 4-1. Low Level Radiation Measurement 112

App. 4-2. Precision Cable Length Measurement 116

Ch. VII

App. 7-1. Data, Applied Physics Laboratory 119

App. 7-2. Data, Jet Propulsion Laboratory 121

App. 7-3. Data, Oak Ridge National Laboratory 123

Ch. IX

App. 9-1. Computer Controlled Satellite Clock 126

APPENDIX 3-1

Peninsula Electronics News, Monday, September 6, 1971 11

Technology Transfer

No-break power systems

Editor's Note: This is the fourth case study in a series on significant government research programs at Lawrence Livermore Laboratory which yield valuable applications in the commercial market. The LLL, operated by the University of California for the AEC, is one of the nation's largest multi-discipline research and development laboratories in the United States. The case studies, which point the way to technology transfer, are described by one of LLL's electronics engineers, Hyman Olken.

By HYMAN OLKEN
Lawrence Livermore Laboratory
University of California
Livermore, California

Our giant, elaborately-interwoven, public utility electrical power systems give us such highly regulated power, so continuously, that we depend on them for many aspects of our daily lives. Thus our subway trains and skyscraper elevators are designed to operate on the assumption that constant-voltage, uninterrupted power will always be coming off the power line network.

a switch is closed to connect our equipment to them. But power outages may occur, and in the case of the LLL research facilities, this may result in much more than mere inconvenience.

For instance, in the new radiochemistry building there are large mass spectrometers in which an extremely high vacuum is maintained by constantly-running, motor driven vacuum pumps. Should power to these pump motors fail, the vacuum chambers would soon become contaminated by air leaking into the system. It may take a week to get the system cleaned up and back to the normal, high-vacuum, operating condition.

Also in this building are large assemblies of electronic equipment for analyzing the radioactive emanations (alpha, beta and gamma rays) produced by the soil samples taken during weapons tests, to determine the isotopes present. These pulse-height analyzer and automatic sample changing systems use many of the most advanced solid state radiation detectors, made of lithium-drifted germanium (GeLi), which have extraordinary resolving power.

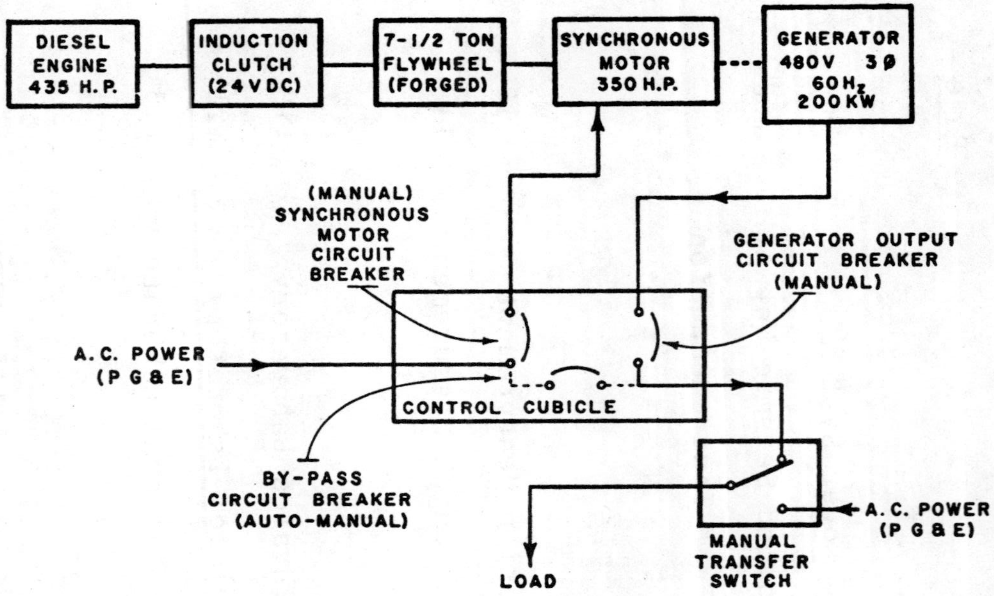

Fig. 1 - Functional Schematic, LLL 200-kW No-Break Power System.

To avoid such disastrous equipment damage and irreparable loss of test data, the Laboratory has provided a standby power generating system which is automatically maintained to energize the critical loads in the radiochemistry building whenever utility-supplied power should fail. Such automatic, standby power generating systems have been in use and commercially produced for a number of years.

One valuable type uses a motor-generator (MG) set in which the motor is driven from the utility power line and the generator output feeds the critical loads. The MG set also drives a flywheel which can be connected, by a special clutch, to a diesel engine. When utility line power fails the energy stored in the flywheel maintains a continuous output from the generator to the critical loads. Also, the clutch becomes engaged and frequency be minimal, and of less than 15-second duration (Fig. 2). In particular, the normal 60-cycle frequency must not drop below 59 cycles. This called for a large (50-inch diameter), heavy (7-ton) flywheel, running at high speed (1800 rpm).

SEVERE OPERATIONAL TOLERANCES

Also, because of this heavy, high speed flywheel it was necessary to add a special device to accurately synchronize the MG set with the utility power line when the MG system was switched back to that line after the power outage was over. If such accurate synchronization were not provided, there would be a current surge of thousands of amperes into the MG set motor, rather than a normal 100- or 200-ampere surge.

that have occurred since.

An equally important advantage of this system is that the isolation furnished by the motor-generator set, and the great inertia of the heavy motor-generator-flywheel mass keeps the output so steady that the power surges and voltage "spikes" which normally occur in the utility system (due to the switching of other loads and generators onto or off the line) are almost completely eliminated (Fig. 2). With the progressive conversion of our radiochemistry building research machinery to solid state electronic equipment, in which the transistors can readily be burned out by line voltages surges, this contributes greatly to eliminating equipment failures. Similarly, the generation of erroneous data (counting) by the counting instrumentation, which may be produced by line voltage spikes, is also eliminated.

Sometimes, although rarely, this assumption does not hold, and the power goes off. Then many people may get marooned in skyscraper elevators, and subway trains may go dead on their tracks. The San Francisco Bay area has suffered very little from this difficulty compared to other metropolitan areas; however, the failure of bulk power is a possibility here as elsewhere.

The highly sophisticated and refined instrumentation systems that are used to carry out the Lawrence Livermore Laboratory's research programs are likewise designed and built on the assumption that plentiful, constant-voltage electrical power will always come flowing off the utility network's lines whenever

LOSS OF POWER MEANS LOSS OF EQUIPMENT

They thus permit detecting a great many of the isotopes present in a sample — something which could not previously be done without elaborate chemical separation of these isotopes. But these highly effective detectors have to work in a vacuum and at cryogenic temperatures. Hence, if the vacuum pumps and cryogenic refrigerators which keep them operating should lose electrical power, these delicate and costly detectors would not only stop operating, they may even become spoiled and not be usable again.

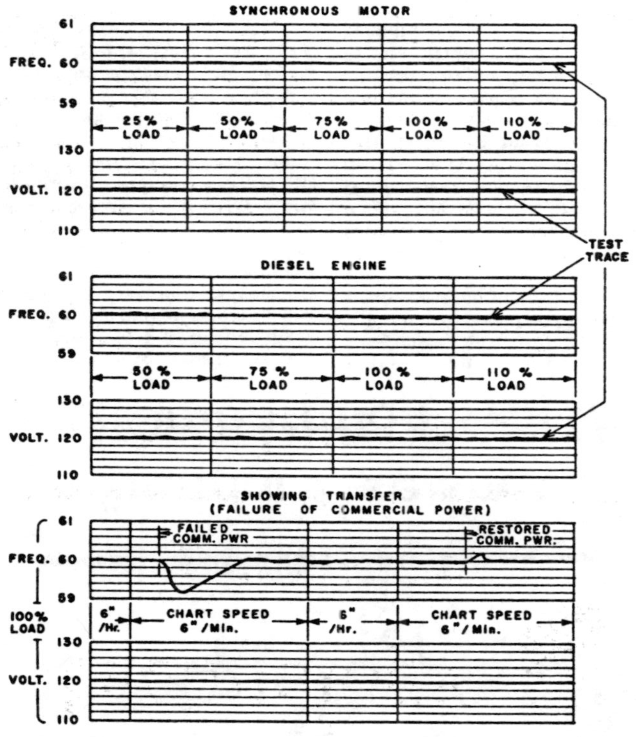

Fig. 2 - Test Record, Operating Characteristics of LLL 200-kW No-Break Power System. (Note exceptionally good voltage and frequency regulation and short transfer time.)

and the flywheel energy is used to crank the diesel engine and get it started. The diesel then becomes the prime mover of the flywheel-motor-and-generator combination (Fig. 1).

Thus the critical loads fed by the generator continue to be supplied, but the energy comes from the diesel instead of the utility-line driven electric motor. When the utility line power comes back on, the MG set is automatically reconnected to the line and the clutch is disengaged, stopping the diesel.

Although manufactured and in use for a number of years, these automatic stand-by power generating systems had not, until recently, been produced in the large capacity (200kW) required by the radiochemistry building. Most of the units produced were of the several-kilowatt, or tens-of-kilowatts size, and a few rare 100-kW units had been produced.

SYCHRONIZATION STOPS CURRENT SURGE

But for the radiochemistry building requirements the standard units could not be merely beefed up to the required size because the critical loads in this building made some exceptionally severe operational tolerances necessary. For example, in many applications where such systems are used it is permissible to take even several minutes to crank up the diesel, hence a small, low energy storage flywheel may be used. Also, it may be permissible for the voltage and frequency of the power supplied to the critical loads to drop appreciably during the switch-over to the diesel. This also reduces the need to store energy in a large flywheel.

But for the radiochemistry building's critical loads, the requirement was that, during the change-over to diesel drive there be no interruption in power flow, and that the changes in voltage

The process of acquiring a system to fit these requirements started with an EE Dept. procurement specification written by W. Abernathy working with Dr. Abe Tilles. Since the flywheel needed to effect smooth transition would be heavy, large and rotating at high speeds, Dr. Tilles, an expert in the EE Dept. who had worked on large power plant turbo-driven generator shafts and flywheels, was called in to make certain that the specification requirements would provide a safe system. This paid off in that our spec called for a single-piece, forged steel flywheel, of proven quality and ductility, in place of the cast steel ones often furnished in similar units.

The finished product suffered some damage in transportation from Wisconsin to Livermore, but was fully repaired and put into operation. As a result it has been in continuous operation in radiochemistry and kept many critical radioactive analysis systems in operation despite several failures of utility-supplied power

GUARDIAN ANGEL HOVERS OVER LINE

Thus the no-break power system is truly a guardian angel that hovers constantly on the line and protects our highly refined, but highly vulnerable radiation analysis instrumentation from the power outages and voltage surges that could easily cause critical operation failures or even destruction of the equipment.

Moreover, as has often happened in the past, to meet the needs of the super-sophisticated and refined research tools required, LLL has to push the development of standard equipment far beyond existing products commercially available. When this is done it not only helps LLL research programs because they get the tools they want, but it also leaves the manufacturer with a much more advanced unit to add to his product line.

Peninsula Electronics News, Monday, September 20, 1971 5

Technology Transfer

'Standard' seismic instrumentation systems

Editor's Note: This is the fifth case study in a series on significant government research programs at Lawrence Livermore Laboratory which are yielding valuable applications in the commercial marketplace. The LLL, operated by the University of California for the AEC, is one of the nation's largest multidiscipline research and development laboratories in the United States. The case studies in this series are described by one of LLL's electronics engineers, Hyman Olken.

By HYMAN OLKEN
Lawrence Livermore Laboratory
University of California

Ever since the development of nuclear explosives, the crucial question has always been, "How powerful an explosion will it actually produce?" The one basic approach for answering this has been, "Set one off and see." But when an explosion equivalent to thousands of tons of TNT occurs, how do you measure it? In the language of the trade, "What's the yield?"

Throughout most of the period of testing in the atmosphere, reliance for measurement of the power of the explosion was based on chemical methods. That is, the fission products of the explosion mary means of determining weapons explosion yield.

To improve this situation, a great deal of development has taken place towards providing radiation detectors of great sensitivity and resolution. As a result, the samples brought up from underground explosions can now be analyzed for their radioactivity merely by putting them into the right detector and measuring their radio-active emissions by the detector output. In the trade this is called "counting."

DETECTORS ELIMINATE CHEMICAL SEPARATION

Prior to development of these advanced detectors it was necessary to first chemically separate the underground samples into their constituent nuclides, then measure each nuclide's radio activity by counting. The new detectors eliminate most of the chemical separation, so the counting can be done as soon as the samples are retrieved. (A detailed explanation of this will be presented in a future article in this series.)

However, this still entails drilling for samples and makes the yield determination a costly, laborious, and time consuming rapidly, through work in the fields of seismology and geophysical exploration for oil, the seismic method of yeild determination looked very feasible.

The Laboratory therefore decided to set up a network of stations which pick up the shock waves in the earth produced by the underground nuclear explosion, and translate them into electrical signals which are then recorded. From these records, by the use of the proper analytical (mathematical) methods, it can be determined how big a shock or explosion took place at the test site.

Specifically, the network consists of four similar stations, one each near Landers, Ca.; Mina, Nev.; Elko, Nev. and Kanab, Utah. At each station our scientists can remain in Livermore and get a report on how big the explosion was almost the instant it occurs. In actual practive, the amplitude of the voltage signals on the recording shocks being picked up; a magnetic tape (specific event) recorder, which runs at high speed to give a detailed picture of each shock; a magnetic tape which runs continuously, to pick

Figure 1 — Control console, Seismic Data Center, Bldg. 111, LLL Livermore. All four seismic sensing stations of the network feed their data to this point, are remotely operated and controlled from this console.

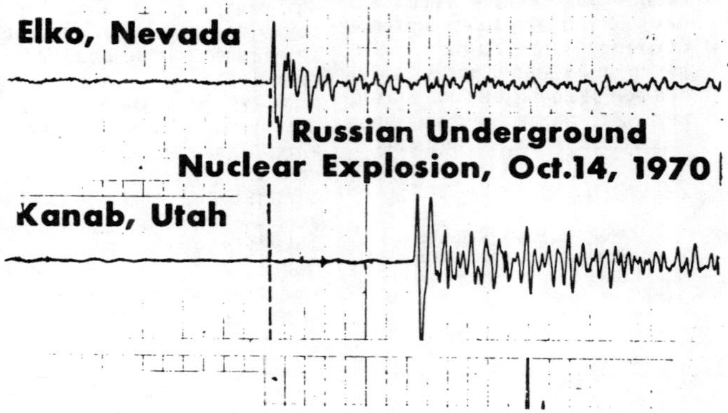

were scraped up, and from these it was determined what nuclear reactions took place and from these, in turn, the power of the explosion could be calculated.

TESTS CONFINED UNDERGROUND

However, when the treaty banning tests in the atmosphere came into effect, tests had to be confined underground, often at great depths. In these cases, getting at the fission products of the nuclear explosion becomes a difficult problem of laboriously drilling holes down to where the explosion took place and bringing up core samples.

Furthermore, such core drilling takes a long time, so the short-life nuclides that would easily be gathered up after an air burst are no longer available. The task of determination of yield by chemical methods therefore became much more complicated and difficult. However, this has been and continues to be the primary operation. What is desired is a method of determining yield that is instantaneous (on line) or close to it; that is inexpensive; and that works equally well for both the small shots (about a kiloton) and the giant shots (in the megaton range). Furthermore, it would be very desirable if the results of the test (the yield) could be reported directly to Livermore as soon as the test occurred, where they could be immediately evaluated.

The possibility of achieving such a method was suggested by the fact that, when an underground shot takes place the explosion generates a shock wave, like an earthquake shock, that spreads out from the point of detonation, throughout the earth, like the spread of a ripple outward from the point where a rock is dropped into a still pond.

Since the technology for detecting and measuring these shock waves in the earth has been well developed for decades past and the methods for analyzing such data have been improving there is a seismometer — a small mass suspended from a frame by suitable springs with the frame mounted on a concrete base. Any vibrations of the earth due to shock waves are transmitted through the base to the frame. Due to its inertia, the mass remains relatively stationary while the frame moves.

SUSPENDED MASS FORMS TRANSDUCER

A coil mounted on the mass is acted on by a magnet on the frame to produce electrical voltages (in the coil). These voltages correspond (over a certain range of frequencies) to the ground motion. The suspended mass and frame thus form a "transducer" which produces electrical signals corresponding to ground motion.

These signals are sent by regular telephone lines to a central recording station at the administration building at Livermore (Figure 1). Hence,

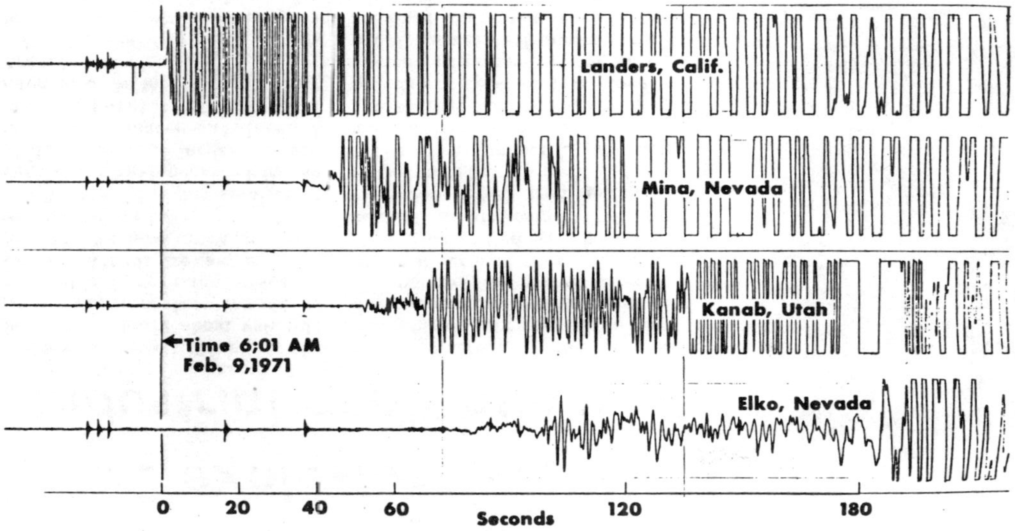

Figure 2 — Record of destructive L.A. earthquake as recorded at Seismic Data Center at instant it happened (6:01 a.m., Feb. 9, 1971). Time difference in arrival of shock (full scale wave amplitude) at different sensing stations permits calculating location of quake center.

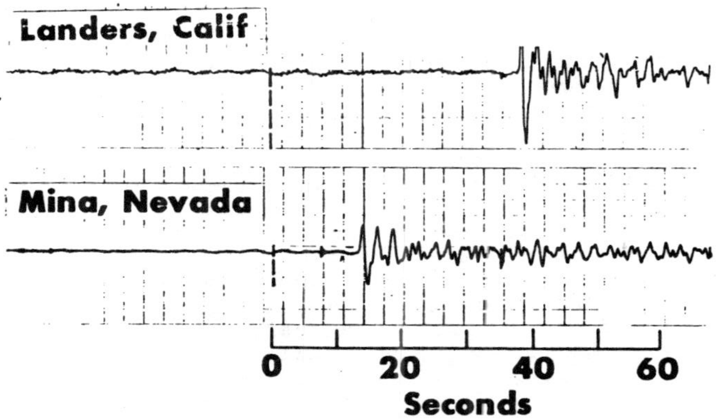

Figure 3 — Record made at Livermore Seismic Data Center of underground nuclear shot so powerful (multi-megaton yield) that its shock wave could be picked up on LLL seismic network after travel from testing ground in Russia.

equipment gives a rough idea, merely by inspection, as to how powerful the explosion was. This is followed by a simple desk calculator computed value also. In the meantime, the shock wave signal data is fed into computers for more precise calculation of the strength of the explosion, or the "yield."

The instrument system which accomplishes this is the result of several years of development work by a group under Otto Krause of EE's Nuclear Apparatus Systems Division. Project responsibility is now under Stan Bishop. From experience they learned, the hard way, that bugs may occur in such a system and how to overcome them so that what was initially a hit-or-miss system now operates every single time, with high reliability.

REFINED STATE OF SYSTEM

The system, in its present refined state, consists of the seismometer unit, which senses the earth shock waves and converts them to electrical signals; a transmission system, for transmitting the shock data to the seismic center at Livermore; a strip chart recorder, which gives a visual record of the up all shocks that occur, but runs at slow speed, to conserve tape (continuous monitoring tape recorder); a remote control system, for operating each sensing station from the Seismic Center at Livermore, and a calibrating system to check the accuracy of the seismometer and transmission and recording systems at each station.

Because of its use in decades past as the primary means of yield determination, the chemical method is still used and is still the one on which primary dependence is placed. However, as experience with the seismometer network is gained, the data can be converted more accurately into yield, and more and more dependence is being placed on the seismic system.

One interesting aspect of this system is that, since it detects and measures shock waves traveling through the earth, it will pick up those caused by earthquakes, or by underground nuclear shots produced in other countries. Both these things were recently brilliantly demonstrated.

In the case of earthquakes, the recent disastrous earthquake in Los Angeles was recorded in full detail at the central station in Livermore, from data picked up

(Continued on page 111)

Peninsula Electronics News, Monday, August 23, 1971

'Packaged' Instrumentation

AEC explosions yield industrial applications

Editor's Note: Perhaps the best way to learn about valuable new commercial products created in Government research programs is via the case method, by studying how much products were developed.

A number of highly significant such cases have occurred in one of the nation's largest multi-discipline research and development laboratories, one that happens to be located in the Bay Area.

This is the Lawrence Livermore Laboratory, operated by the University of California for the AEC. PEN will feature cases that point the way to technology transfer, described by one of LLL's electronics engineers, Hyman Olken.

By HYMAN OLKEN

Lawrence Livermore Laboratory
University of California
Livermore, California

Even during the development of the atomic bomb scientists realized the extremely compact and powerful explosive would have valuable commercial applications.

Hence for many years the AEC's development of nuclear weapons has been paralleled by a "Plowshare" program for application of nuclear explosives to non-military, industrial functions. Concepts have been developed to use nuclear explosives to build canals, to dig harbors and to cause deep underground explosions. These deep underground explosions would fracture a great deal of the underground rock, thus making it porous releasing large underground pockets of gas and signals from the related instruments are run from the nuclear explosive in the hole, up to a point on the ground directly above it (ground zero). The hole is then backfilled with rock and cement to prevent any of the explosion effects from escaping to the surface.

The basic principle of such nuclear explosions for creating new sources of oil and natural gas sounds simple — merely place a nuclear explosive underground, fire it, then tap off the upcoming gas or oil. In actual practice this becomes a complex and difficult operation, for several reasons.

At great depths the ground temperatures may be high enough to damage the nuclear explosive or hamper its operation. Therefore, cooling equipment must be included in the canister which holds the explosive. Also, the instrumentation needs electrical power for its operation, so this must be furnished from above ground. And, of course, there must be many signal and safety devices not only to produce arming and firing of the explosive when desired, but also to ensure that it is not accidentally set off before that time.

Finally, to conserve costs for industrial detonations, the many separate cables used in military tests for monitoring the blast, are combined in a single cable which runs from the explosive below ground to a control or relay station directly above it at ground zero.

This relay station is housed in a container unit which is easily movable from place to place. The unit is connected by a radio link

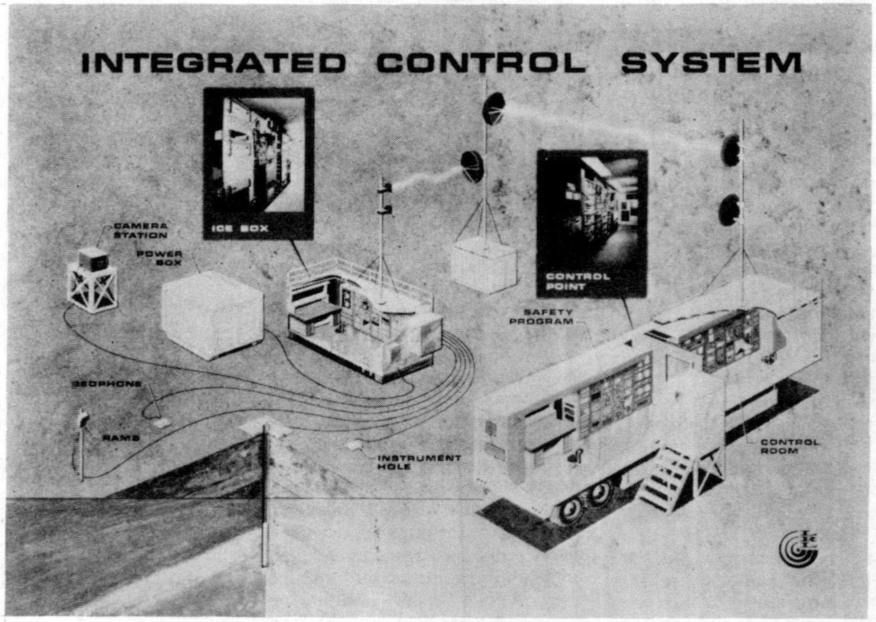

FIGURE 1

sary electronic equipment for a particular event. It is powered from the Power Box, an 8x8x15-ft. "transportainer" outfitted with two identical diesel generators. One generator supplies the power. If it fails, the other circuit is switched on automatically.

The ICE Box is the ground-zero terminal of the microwave link. It distributes the commands downhole and to close-in surface installations, provides all downhole power, and transmits to the Control Room the downhole monitors (signals for monitoring operations). The ICE Box also provides all necessary pre- and post-shot safety information such

trailer is available from the local power system at the control point, or from trailer-mounted diesel engine driven power systems commonly used by the EE Dept. for field tests. Standby power is included to maintain communication with the ICE Box and the outside world in the event of local power failure. The detonation could be conducted and safety data monitored for several hours — sufficient time to restore power or establish other safety controls if needed.

ENTIRE ICS SYSTEM MOBILE

field by shipping a completely checked-out system after all site preparations are completed and accepted.

Thus the know-how developed by EE Dept. engineers and technicians in developing the instrumentation for testing military nuclear explosives has been "cashed in" on and applied to the successful achievement of low cost, safe, transportable, and easily usable equipment for instrumentation and control of underground nuclear explosives. This opens up the widespread use of nuclear detonations for industrial recovery of valuable minerals in the depths of the earth derground oil and gas deposits that would otherwise remain locked up in the rock formation.

Lately, enough progress has been made in the Plowshare studies to allow actual application projects. Instrumentation and control facilities for these projects are required to provide information concerning the underground conditions where the explosive will be detonated, and check-out (monitoring) information to ensure that all equipment and environmental conditions are correct to ensure a safe and effective detonation.

INSTRUMENTATION AND CONTROL REQUIREMENTS

From their experience with the instrumentation and control facilities for developing the military nuclear explosives, engineers of the EE Dept. of the Lawrence Livermore Laboratory (formerly Lawrence Radiation Laboratory) were able to develop a standard instrumentation and control system for Plowshare applications which meets these basic commercial requirements:

1. It is easily "packageable" and therefore easily movable to distant locations.

2. It requires very little manpower to operate.

3. It is very safe.

4. It vastly reduces the time for setting up and constructing an underground test.

To see this system in perspective, we must first take a brief look at what is involved, physically, in one of these contained, underground operations.

Such underground detonations are accomplished by drilling a hole to the desired depth (as for an oil well). A nuclear explosive device, housed in a canister which contains necessary electrical and electronic equipment for firing the nuclear explosive, is lowered to the bottom of the hole while cables to carry the necessary firing signals to a control point (CP) at a safe distance away where observation of the test data and facilities for control of the test are concentrated. Because of the radio link the control station can be located a considerable distance away from ground zero without the use of many thousands of feet of expensive cable.

STANDARD, COMMERCIAL INSTRUMENTATION AND CONTROL SYSTEM

After a few years of development, a system that meets these requirements is ready for operation. The complete system, shown in Fig. 1, is called the "Integrated Control System," which has these construction and design features.

Three major elements make up the Integrated Control System: Control Room, Integrated Control Element (ICE) Box, and the equipment packaged in the explosive canister. Communication between the control room and the ICE box at ground zero is by microwave radio, and from the ICE Box to the explosive canister by coaxial cable. Ancillary equipment, such as a basic power supply at ground zero and a microwave repeater station, are also included in the complete system. These units and their functional relationship are shown in Figure 1.

The ICE Box, an 8x8x20 ft., air-conditioned, commercial cargo container, houses all the necessary equipment such as weather, TV surveillance, radiation area monitoring, and geophone data.

For the fully-contained detonation, the ICE Box will be located no more than a few hundred feet from the emplacement hole (Figure 1). Shock-loading of the box during the ground motion will be mitigated by crushable foam pads, reducing the accelerations to about 5 g's maximum, a safe limit for the equipment inside.

The explosive canister contains a command and monitor system, as well as power supplies for the various subsystems involved in the firing of the explosive. A coaxial cable carries power and all information between the ICE Box and the explosive canister. Power is controlled in the ICE Box by commands from the Control Room.

The Control Room includes the following items:

1. **Control console**
2. **Monitor displays (lights and strip charts)**
3. **Clocks showing world time and countdown time**
4. **Microwave equipment**
5. **Safety program data displays (weather, radiation, ground motion, etc.)**
6. **Communications to Test Manager, etc.**
7. **Standby power system**

The control room is built into a trailer (Figure 1) that includes air-conditioning equipment and provides mountings for the radio antenna. Basic power for the entire ICS system is fully mobile, and completely set up and checked out at some well-equipped location such as LLL-Livermore before shipment to the detonation site. Preparations for shipment do not require disassembly except for radio antennae.

To summarize, the ICS (Integrated Control System) approach for fielding Plowshare events presents the first major departure from the routine weapons test procedures. It allows more efficient use of manpower by combining all functions under a single control, a minimum fielding time, and hopefully, elimination of technical delays in the

Those interested in further particulars of this system will find them in the UCRL report 72148 on "Integrated Control Systems for Nuclear Explosives" by W. F. Ragsdale, which he presented at the American Nuclear Society Meeting on "Engineering With Explosives" held at Las Vegas in January 1970. For a good background on the application of nuclear explosives for recovering underground resources, see UCRL 14756 on "Industrial Applications of Contained Nuclear Explosions" by Donald E. Rawson. (Order UCRL reports from National Technical Information Service, Springfield, Va. 22151.)

SEISMIC INSTRUMENTATION SYSTEMS

(Continued from App. 3-2, p. 109)

by the four stations of our network (Figure 2). By making use of the different time delays in arrival of the signals at the four stations it was possible to compute how long it took the earthquake shock to travel to each station. From this information it was possible, by triangulation, to locate the exact center of the Los Angeles quake.

In the case of picking up underground nuclear explosions from other countries, one such event occurred on the night of October 14, 1970, when our seismic network recorded an earth shockwave (Figure 3). Analysis of the data by U.S. scientists told them that it was caused by a gigantic (multimegaton) shot set off by the Russians.

Thus this seismic network, in reality a stethoscope to pick up shock waves in the earth, is a valuable addition to our diagnostic tools for nuclear weapons development, and provides valuable byproducts in the detection of nuclear shots by other countries and in the study of earthquake phenomena.

In view of the rapidly growing and increasingly widespread interest in earthquake hazards, it is very likely that some of our states and some foreign countries will install similar seismic sensing and recording systems, to further the study of earthquake prediction and earthquake-resistant structure design. Because of this system's use of regular telephone lines for transmission and its other practical features, it may well become the standard design that many others will follow. (Further information about the system may be obtained from the U.S. Atomic Energy Commission, Division of Technical Information Extension, P.O. Box 62, Oak Ridge, Tenn. 37830.)

Technology Transfer

Electronics, radiochemistry

Advanced nuclear rad

(Editor's Note: The following is the seventh case study in a series concerning government research programs conducted at Lawrence Livermore Laboratory which yield important applications in the commercial marketplace. The case studies in this series are described by one of LLL's electronics engineers, Hyman Olken.)

By HYMAN OLKEN
Lawrence Livermore Laboratory
University of California

One of the astonishing discoveries that ushered in the nuclear age was the fact that some heavy atoms of complex nuclear structure spontaneously threw off a part of the nucleus (alpha particle) to decay to a new "daughter" element.

This decay can take place in a number of successive steps to form a complex decay "chain" or "series" which ends up with an element of stable nuclear structure such as lead. Furthermore, each step in the decay chain is often accompanied by other emissions along with the alpha particles, such as electrons (beta-rays), and the photons of energy in the form of electromagnetic rays (mainly gamma rays — a form of very hard X-rays).

This opened the door to a wholly new branch of chemistry — one in which you detect the nature and amount of the element present by detecting the emissions (alpha, beta, and gamma rays) coming out of a sample of material rather than by the classic "wet" chemistry of studying the reactions that take place during the nuclear explosion and thereby point the way to better design of nuclear explosives.

But implementing this tool has required development of a whole vast technology of instruments for detecting and measuring the "emanations" (alpha, beta and gamma rays) produced as part of the nuclear decay or disintegration process. Gamma radiation is the most widely applicable in characterizing radio-nuclides, hence the greatest area of development has been in gamma ray detection and measuring devices. Although alpha and beta detection and measurement are also important, to keep this article brief we will highlight the chain of developments in gamma detection and measurement only.

'IONIZATION AVALANCHE'

The first approach to detecting and measuring gamma rays was to utilize their effect in ionizing a gas. Like X-rays, gamma rays are very penetrating, so the first detectors were metal (or glass) tubes in which was enclosed the gas to be ionized. By proper choice of the gas and proper application of electrical voltages, the gas inside the tube, when ionized by the incoming radiation, could be made to produce an electrical breakdown or "ionization avalanche" inside the metal tube, thereby generating an electrical pulse.

therefore, to produce detectors which not only increased the frequency of their electrical pulses (pulse rate) with the increasing intensity of the radiation; it was necessary also to produce detectors which produced larger electrical (voltage) pulses the higher the photon energy (or frequency) of the incoming radiation. This could be done by appropriate change in the electrical characteristics of the Geiger-Muller tube, and these were appropriately called "proportional" counters. That is, the output pulse height (amplitude) was proportional to the energy of the incoming gamma photon.

If one then arranged an electronic "discriminator" circuit that would separate the pulses of different height (i.e. amplitude), so they could be counted separately, the speed at which these different counts accumulate would indicate the diagnosticians only if they could isolate one radioactive element from the entire chain of isotopes that were forming simultaneously, and accumulating on top of one another, during the radioactive decay scheme (process).

SINGLE CHANNEL RADIATION MEASUREMENT

This "single channel" radiation measurement technique therefore required a lot of "wet" chemistry methods to separate the one element (Isotope) to be measured from all the others in the mixture of elements being formed during the decay series.

Although this technique was useful, the need for wet chemistry preparation made its was preceded by the development of the sodium iodide (NaI) detector, in which the nuclear radiations generate pulses of visible light as they traverse a sodium-iodide crystal. These light pulses are picked up (by a photocell) and converted to electrical pulses which are then fed to a pulse-height-analyzer. The analyzer converts these electrical pulses, as it does those from a semiconductor detector, into a curve (spectrum) of the incident nuclear radiation.

The NaI detector came into wide use because of its great sensitivity. However, its resolution is quite poor, so the spectrum obtained does not have the sharp peaks required to achieve practical radiochemical analysis (Fig. 1).

With these high resolution detectors and the high speed formation of the whole gamma spectrum on a CRT screen, the radioactive analysis technique

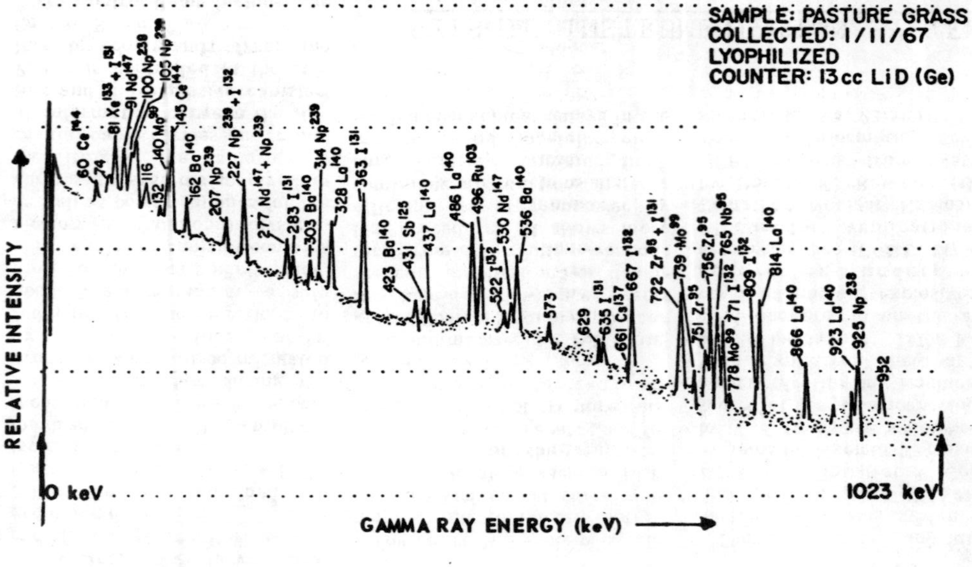

place between different elements in solution, during oxidation, etc.

NEW BRANCH OF CHEMISTRY DEVELOPED

This new branch of chemistry, appropriately called "radiochemistry," was just the tool needed for studying the effects of nuclear weapons. That is, samples of the radioactive material picked up after a test shot could be "analyzed" to determine what radioactive elements (isotopes) they contain and how much of each. This could shed a great deal of light on what

Furthermore, the greater the intensity of the incoming radiation, the greater the frequency of the pulses. Even the earliest of these detectors (Geiger-Muller tubes) could produce thousands of pulses per second in response to sufficient incoming radiation, so electronics engineers developed "scalers," or high speed counters, to indicate the rate of generation of the electrical pulses, and thereby the intensity of the radiation.

But to do this radioactive analysis effectively, it is necessary to detect and measure the whole group of gamma rays, of different energies, that occur during each stage of nuclear decay. It was necessary,

Figure 2 — Local pasture grass contaminated by Chinese bomb fallout. Spectrum was produced from a 13 cc planar Ge(Li) detector.

intensities of the different energy level gamma rays present in the incident radiation. If one plotted each intensity against the corresponding photon energy level, this would produce a panoramic picture or "spectrum" of the gamma rays being emitted during that stage of radioactive decay.

This is very helpful — provided you are "analyzing" only one radioactive element at a time. Hence this "single channel" method of radiation measurement to detect and measure the quantity of a radioactive element could be used by our application laborious and difficult, so electronics engineers evolved the next stage of instrumentation known as the "multichannel" analyzer.

Instead of measuring, one at a time, the pulse rates for each of the different gamma energies in the incoming radiation, the pulse rates for all gamma energies were measured simultaneously, but the pulse rate for each gamma energy level is measured in a separate "channel," and the results stored in a separate "bin" or location in a core memory.

Digital logic circuits were provided so that these different rates could be read out sequentially to form a histogram on a cathode ray tube. This histogram is, in effect, a picture of the spectrum of gamma radiations of different energies coming out of the radioactive sample. This device, a multi-channel pulse-height-analyzer, in effect produces for the experimenter a continuous picture of the whole spectrum of gamma radiations coming from the sample, instead of his having to plot it laboriously from the data produced by the single-channel analyzer system.

This advance was further supplemented, in the last several years, by the evolution of a new semiconductor type of radiation detectors. These detectors, which are made by drifting lithium into a single crystal of germanium, produce electrical pulses which correspond very accurately, in amplitude (voltage), to the energy of the incoming gamma radiation. The output pulses of these semiconductor detectors correspond to the energies of the incoming gamma photons far more accurately (Fig. 1) than did earlier types of detectors. This is called technically "high gamma energy resolution."

The semiconductor detector becomes so sensitive that it is generally no longer necessary to separate out each individual radioactive element from the sample by wet chemical methods and measuring its radiation output separately.

Instead, all the radiations from a sample containing a complex mixture of radioactive elements can be measured at once and all the measurements can be displayed simultaneously as a curve on a cathode ray tube screen, which curve will have a large number of sharp peaks, each one corresponding precisely to a different gamma energy.

CHINESE FALLOUT DETECTED EASILY

It is then possible to pick out the radioactive elements present in the sample, and the amount of each one, merely by inspection of the sample's spectrum on the CRT screen. This is well illustrated by the picture (Fig. 2) of the isotopes present in the sample of grass contaminated by Chinese fallout.

(See "Gamma-Ray Spectrometers for the Assay of Complex Mixtures of Low Concentration of Radionuclides in Environmental and Biological Materials," Paul L. Phelps. IEEE Transactions on Nuclear Science, Feb. 1968, Vol. NS-15 No. 1, pp 376-382. Also available, with large, clear illustrations, from the National Technical Information Service, Springfield, Va. 22151, as UCRL 70544.)

The greater the number of channels in this system, the greater the precision of the results, and the more accurately the contents of the sample can be analyzed. So engineers have

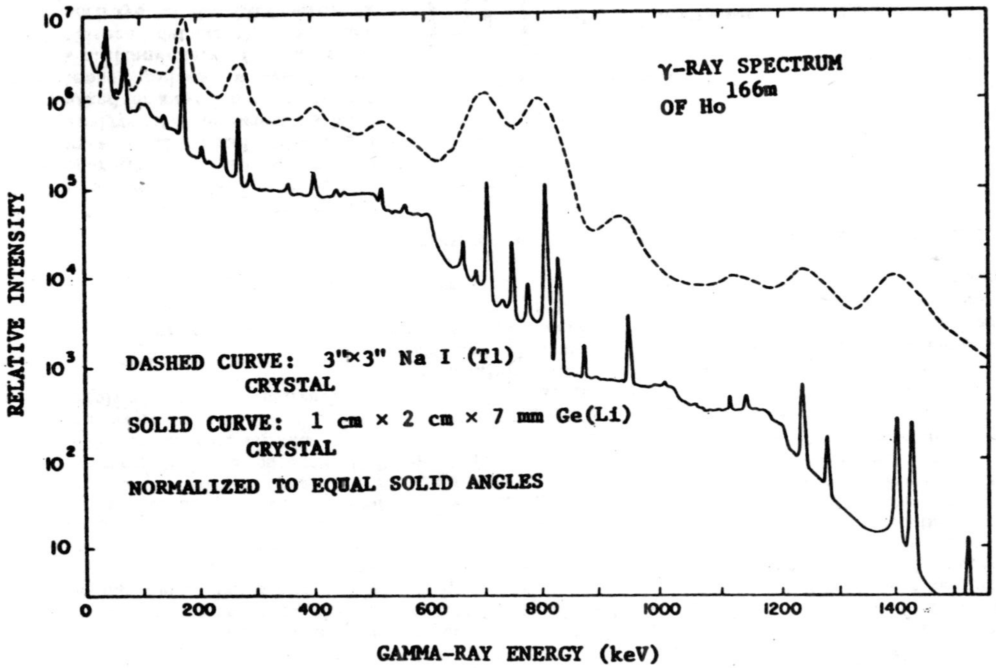

Figure 1 — Improved resolution of gamma ray spectrum of Ho 166m by the new germanium detectors over the resolution achieved with sodium iodide detectors. The curve obtained with the NaI crystal shows only partial separation of the gamma energies. Thus two partially resolved peaks in the 700 to 800 keV region of the NaI spectrum are changed to eight well-resolved peaks, corresponding to eight distinct gamma rays.

(Cont. p. 114)

iation measurement tools

pushed the development of multi-channel analyzers from 100-channel analyzers to 256-channel in the middle 1950's and now to the 4096-channel type. This requires more logic circuits and larger core memories, until now it is profitable to use a small computer and a large core memory to do the job, instead of using fixed logic circuits in the conventional pulse-height analyzer system configuration.

during the lithium drifting process and hence very few of the detectors produced had the desired characteristics.

A development program was therefore undertaken a few years ago by the Research Division of the EE Dept., and much of the original technology of fabricating these detectors was developed here at LLL. This program led to the development of standards and processes for producing germanium of the extra-ordinarily high purity required to make the detector manufacturing (lithium drifting) process work.

To these standards and processes this group at LLL

ation, particularly those which occur at extremely low radiation levels. To attain this potential field of nuclear research it was necessary to develop not only this technology of detector manufacture, but also the technology of eliminating the stray radiation that crops up from many sources and makes precise measurement of nuclear radiation at very low levels extremely difficult.

This field of technology was developed by another EE Dept. group, this one headed by Paul L. Phelps. The resulting low level radiation measurement technology (Fig. 3) has set such high levels of precision measurement in the study of the physiological

COMPUTER INCREASES PULSE SORTING

But the computer makes possible far more pulse sorting operations than just those needed in a multi-channel pulse-height analyzer. So the scientists have asked for a system which detects those pulses which occur simultaneously (using coincidence detectors), and then portrays on the CRT screen a number of curves simultaneously, each of which curves corresponds to a different gamma-emission spectrum.

Thus, each curve that appears corresponds to the gamma emission spectrum in coincidence with a different gamma ray. With these "3-D" spectrum analyzers it is possible to tell, not only the different radioactive elements present in a sample; it is also possible to analyze the steps of the process that occurs in one stage of the decay scheme, or the disintegrations of one element into another. With this capacity to "see into" the decay process, the way has been opened to a vast new area of advance in the study of the structure and decay processes of the atomic nucleus.

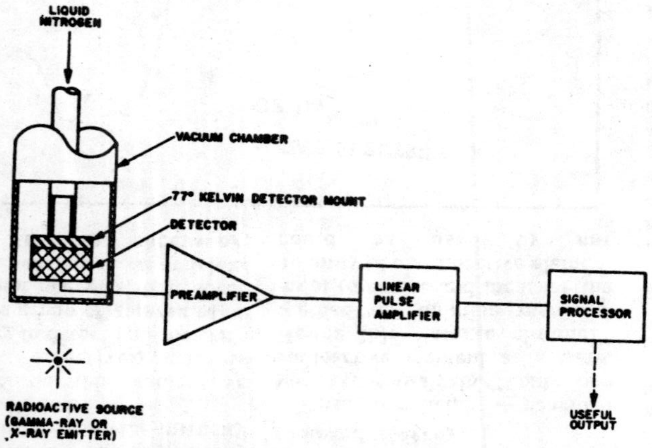

Figure 3a — Basic scheme for nuclear radiation measurement. Radiation converts incident radiation into electrical pulses. Rate of pulse generation, measured by electronic (counting) devices indicates intensity of the radiation.

made substantial contributions, notably those of Guy Armantrout, and they have helped also in getting these standards and processes adopted industry wide. These semiconductor radiation detectors are therefore now

effects of radiation that it has been studied and copied widely by other laboratories. Associated with the EE Dept. group in the development of this low level radiation measurement technology were a group from the

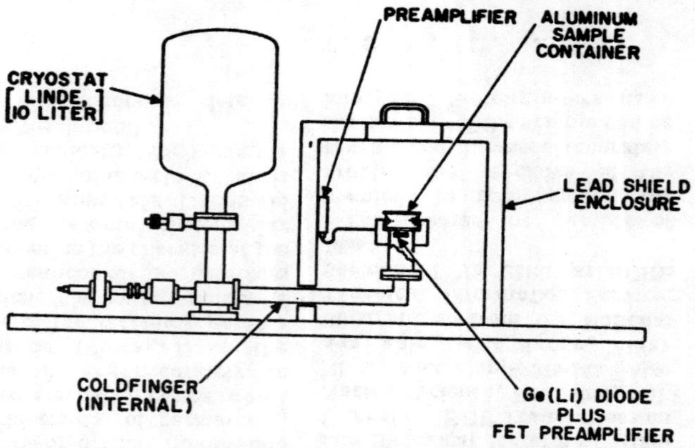

Figure 3b — Physical arrangement for semiconductor detector nuclear radiation measurement. To attain desired operating characteristics detector must be enclosed in a vacuum chamber, kept cooled to cryogenic temperatures (-77 degrees C), and shielded from stray radiation.

available from the National Technical Information Service, Springfield, Va. 22151.)

UCRL 73001, "Principles of Operation of Semiconductor Detectors," by Guy Armantrout.

UCRL 73019, "Fundamentals of Semiconductor Detector Systems Design," by A. E. Bradley.

UCRL 70544, by Paul Phelps, cited earlier.

UCRL 72108, "Ge Crystal Growth and Evaluation as Ge(Li) Detector Material," by R. Wichner, G. A. Armantrout and T. G. Brown.

However, to achieve these promising new fields of nuclear research required first developing the technology of manufacturing these high-resolution semiconductor detectors. For it was found that the raw material — germanium — behaved in unpredictable ways commercially obtainable in the shapes, sizes, and performance characteristics required for almost every research application.

Among the great potential areas for nuclear research opened up by these high resolution detectors is the study of the biological effects of nuclear radi- LLL Mechanical Engineering Department who developed the associated mechanical equipment.

To learn more about these advanced developments in nuclear radiation measurement technology, see the following Lawrence Livermore Laboratory (UCRL) reports. (Copies are

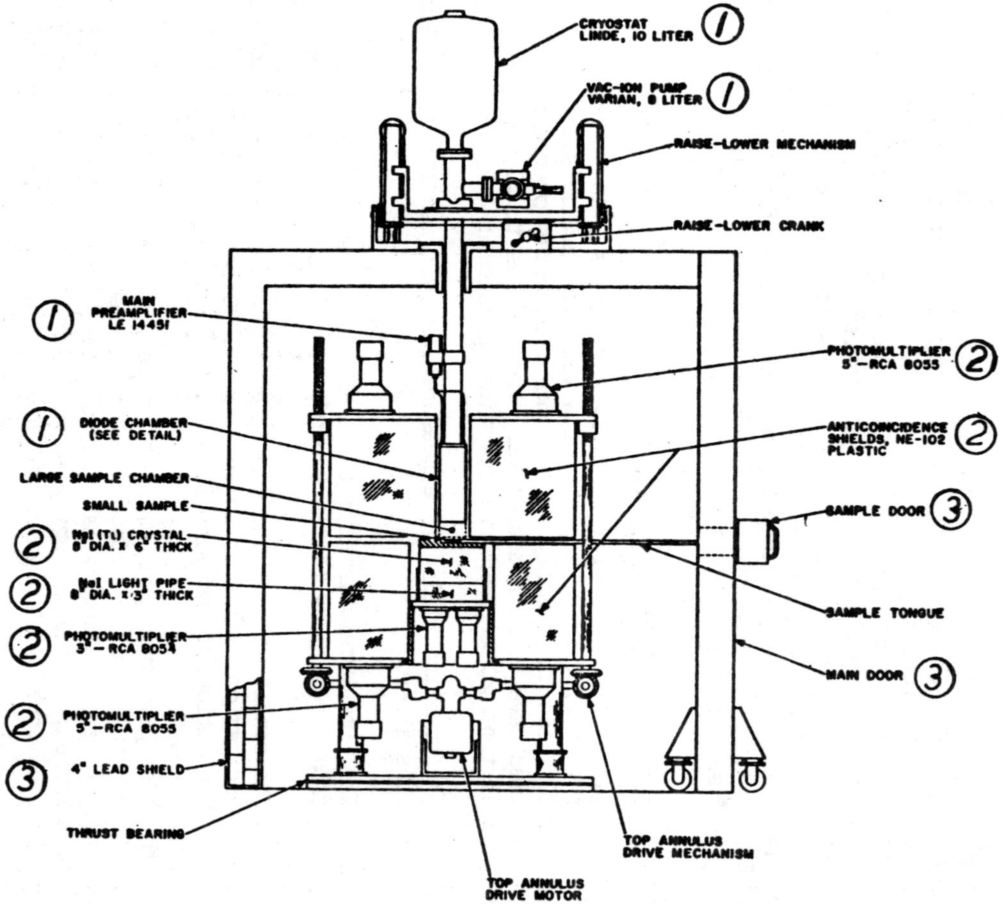

Figure 3 — Low-Level Nuclear Radiation Counting Technology. The study of biological effects of radiation requires technology for ultra-low level radiation measurement which comprises (1) Sensitive, high resolution, Ge(Li) radiation detector, enclosed in a vacuum chamber and kept at liquid nitrogen temperature for best operating characteristics. (2) Anti-coincidence measurement arrangement, which eliminates secondary radiation produced by primary radiation inside the measurement apparatus. (3) Thick metal shield encloses the measurement system to eliminate natural radiation background present everywhere.

ELECTRONICS ENGINEERING DATA LETTER

ELECTRONICS ENGINEERING DEPT.
LAWRENCE RADIATION LABORATORY
LIVERMORE, CALIFORNIA 94550

August 1968 VOL. XII. No. 8

TIME DOMAIN REFLECTOMETRY
EE DEPARTMENT DEVELOPES STATE-OF-THE-ART INSTRUMENTATION FOR IT

One of the great achievements of the builders of the first trans-Atlantic cables was to devise a method of making electrical measurements, at one land end of the cable or the other, which would determine at what point along the cable a break in the circuit had occurred. This saved vast amounts of money in locating the cable break and repairing it.

In this atomic age, we use different kinds of cables (coaxial) for different purposes (fast pulse transmission from device shot instrumentation). However, every shot involves a huge array of cables to connect the necessary detectors to the measurement and recording devices. Often it is imperative to know precisely how long it takes a pulse to travel over a given length of cable. It is also often important to know precisely where an electrical fault exists along the cable. Using some of the electrical measurement principles first developed by the trans-Atlantic cable trouble shooters, the art of determining a cable's characteristics by making electrical measurements at one end of it has flourished into a whole branch of electrical technology called "time domain reflectometry". This field has long been a major field of work for the EE Dept. and the department has made a chain of notable advances in developing the instrumentation equipment required for it.

What makes "time domain reflectometry" particularly difficult, especially in those applications of concern to LRL, is the fact that the velocity with which an electrical pulse travels along a cable is a large fraction (60% to 92%) of the velocity of light, which velocity is one foot per nanosecond (one billionth of a second). Since the cables over which pulse travel times have to be measured vary from inches to 10,000 feet, the pulse travel time measurement creates these problems:

1) The time intervals to be measured are very short (microseconds or smaller).

2) The time intervals have to be measured over a wide range - from picoseconds (10^{-12} seconds) to microseconds (10^{-6} seconds).

3) The sensitivity of the time measurements has to permit clear indication of 1/4 of an inch in 10,000 feet of cable, or a resolution of one part in 500,000!

To cope with these problems, progressively better pulse travel time measurement techniques were developed. The first used the standard laboratory technique of inserting calibrated delay elements, in place of the cable to be measured, in the measuring--circuit until signals were matched on an oscilloscope screen. A significant step forward was made with the

"Three Oscilloscope" method, in which the cable was left continuously in the measuring circuit and both the pulse sent down the cable and its reflection, after it returned from the open end of the cable, were shown as peaks on an oscilloscope screen, a few inches apart. The actual measurement was made by filming the oscilloscope screen and measuring the distance on it between the start and reflected pulses. Though more direct, this method was still slow--it took two or three weeks to complete a cable network measurement; also data reduction from the films had to be done by experts by hand, and it required a rack full of equipment. (A brief description of these two methods is given in the UCRL report No. 7078, by E.F. Laine, entitled A Method of Measuring Pulse Delay in Coaxial Cables to Nanosecond Resolution.)

The concept which paved the way to really successful pulse travel time measurement was the idea that, instead of using the start and reflected pulses to show up as peaks on an oscilloscope, they be used to start and stop a flow of high frequency (100 megacycle) pulses generated by a local oscillator. The number of pulses passing through between the start and stop signals could be counted on the newly available high-speed electronic counters, thus giving a very precise, digital measurement of the time it took the pulse to travel to the end of the cable and be reflected back. (This scheme is also briefly described in UCRL 7078.)

This sequence of operations was made automatic by a one-tube circuit arrangement devised by Ed Laine, which Bob Werner further developed so that it gave a direct, visual reading of the desired cable length (one way pulse travel time over the cable being measured) in nanoseconds. Werner later developed a small transistorized version of the same unit. This made available a "cable calibrator" device which gave immediate, automatic cable length measurement readings, to the accuracy (resolution) desired, and which was available in a small handy package. UCRL 7078, mentioned above, gives full details of this development. A self-contained field-use instrument, with its own power supply and counter, was developed by Tom Holdsworth and John Glasgow (LE 13807-1).

To bring the cable length (pulse travel time) measurements up to the range and resolution required, it is often necessary to use a "bias" cable in series with the cable to be measured. After the measurement of the unknown plus bias cable is made, the known length of the bias cable has to be subtracted from it, to give the actual length of the cable being measured. For cable measurements made in the field this was a nuisance, so the "cable calibrator" was improved so that it would do this subtraction automatically, by adding to it a miniature computer and a memory.

With this unit ("cable calibrator, auto subtract," LE 15313) any technicain, with a few minutes training, can instantly make the measurements which formerly required an engineer and considerable time to do. Also this unit costs $3000 compared to the $11,000 for the initial cable length measuring equipment. So successful has this system been that it is now in wide use by most AEC contractors (Sandia, EG&G). This development was made by John Glasgow of the Long Range Diagnostics Development Group, under Gordon Longerbeam, in the EE Department's Field Test Systems Division.

APPENDIX 4-2

Having such an automated cable length measuring device, EE engineers went one step further and produced a device (for Plowshare Program) which makes continuous measurements on the length of a single cable. Before the nuclear device test this cable is placed so that the chimney, produced by the underground device explosion, will cut part of it off. The part of the cable left intact and operating is thus measured, and a continuous record of its length gives, in effect, a time curve of how far the chimney spread. With a number of such cables placed in strategic spots and their lengths measured continuously during the shot, it becomes possible to work up a diagram which "profiles" the configuration of the chimney produced by the device explosion, as the chimney developed. The instrument which makes these automatic measurements, LEA67-20837, is called the "cliper", an acronym for "Cavity-Collapse Location Indication by Pulsed Electromagnetic Reflection" and was also developed by the same EE group. This system, also designed by John Glasgow of LRDD, was conceived by Ray Stefani of Plowshare Program Group of the Nuclear Apparatus Systems Division of the EE Department, and Clyde Sisemore,a Plowshare Physicist.

Thus the end product of this long trail of development--from the trans-Atlantic cable pioneers to the EE engineers devising sophisticated instrumentation for diagnosing the effects of nuclear device explosions--is a mechanism which can give the scientists a "picture" of the cavity produced underground by the explosion of a nuclear device. This is extremely valuable to the scientists of the Plowshare program, in giving them a "picture" of the rock-shattering effect produced by the nuclear explosion, and thus providing a key tool for developing underground mining methods by use of nuclear devices.

HO. GL

cb

Distribution:
Engineers
Supervisors
Coordinators & Designers
Technicians
Draftsmen
Patent Engineering

(1) NAME: Applied Physics Laboratory, The Johns Hopkins University (APL-JHU) **DATE: October 13, 1967**

(2) LOCATION(S): 8621 Georgia Avenue, Silver Spring, Maryland 20910, Main Laboratory complex in Howard County, Maryland, near Scaggsville

(3) SPONSORING AGENCY: U. S. Navy, Naval Ordnance Systems Command

(4) MANAGEMENT ORGANIZATION: The Johns Hopkins University

(5) CHIEF ADMINISTRATOR: Dr. Ralph E. Gibson, Director

(6) CUSTOMERS SERVED: FY 1967: Navy (80%), NASA (7%), DOD (3%), Army (2%), Air Force (7%), NIH (*%), ESSA/Dept. Com. (1%)

(7) TABLE OF BUDGET AND PERSONNEL:

Year (FY ending 6/30)	1942	1943	1944	1945	1946	1947	1948	1949	1950	1951	1952	1953	1954
Total Budget ($ Million)	0.3	2.5	3.6	4.8	4.5	2.8	1.7	3.9	4.4	6.0	8.5	8.9	10.1
R & D Budget ($ Million)					Same as Total								
Total Personnel	269	491	650	886	684**	753**	754**	682	698	818	932	1041	1101

(Continued)

Year (FY ending 6/30)	1955	1956	1957	1958	1959	1960	1961	1962	1963	1964	1965	1966	1967
Total Budget ($ Million)	11.3	11.7	14.9	14.3	18.4	22.8	31.1	39.5	47.7	49.3	47.3	46.3	62.9
R & D Budget ($ Million)					Same as Total								
Total Personnel	1172	1247	1329	1388	1564	1735	1954	2218	2450	2447	2467	2556	2564 Est.

(8) ORIGIN OF CENTER: In 1940 the National Defense Research Committee of OSRD was faced with an urgent problem of defending the fleet against air attack. A group of scientists under the general direction of Dr. Vannevar Bush and Dr. R. C. Tolman, and immediate leadership of Dr. M. A. Tuve, Dr. L. R. Hafstad, and Dr. R. B. Roberts of the Carnegie Institution of Washington studied problems of improving anti-aircraft capability. In 1942 Dr. Bush asked Johns Hopkins to manage a laboratory to accelerate development of an anti-aircraft shell proximity fuse designed by the Carnegie Institution. Operations began in April 1942 under direction of Dr. Tuve and production began in late 1942. First combat use in January 1943 successfully downed enemy planes. Original location was in an available building in Silver Spring, convenient to JHU and C.I. of W.

(9) ORIGINAL AND SUBSEQUENT MISSIONS: Original mission to develop proximity fuses continued throughout World War II. In 1943, APL undertook development of an anti-aircraft gun director, made operational in 1944. In late 1944, when APL's sponsorship shifted from OSRD to the Navy, the Bureau of Ordnance assigned it the mission of R & D in guided missile technology and responsibility for developing missiles for fleet use. The Bumblebee program has involved the sequential development of operational missiles and technological innovations for the guided missile program, including launching, propulsion, control, guidance and warhead weaponry. In 1958, the Navy and ARPA began sponsoring investigations in satellite navigation systems. APL's major program categories (as of FY 1967 with percentages of staff assigned to each) are: (a) Naval surface-launched missiles and missile systems (26%); (b) Guided missile supporting R & D (13%); (c) Ballistic missile systems (14%); (d) Navy Satellite Navigation Systems technology (24%); (e) NASA space technology: satellite for geophysical research; life-support; and Gemini/Apollo communications systems (7%); (f) Theoretical

App. 7-1 — Data Summary, Applied Physics Laboratory, major multidiscipline R&D laboratory operated by Johns Hopkins University for U.S. Navy Department and Experimental Exploratory Studies (3%); (g) Fundamental research, self-initiated (6%); (h) Special War Gaming Studies for Chief of Naval Operations (2%); and (i) Miscellaneous projects (5%).

(10) MAJOR ACCOMPLISHMENTS: (a) Development of proximity fuse during WWII contributed greatly to Naval successes in the Pacific, Army artillery effectiveness, and defense against Nazi V-1 missiles; (b) first supersonic ramjet flown by APL in June 1945; (c) control at supersonic speeds in 1947 and fully guided supersonic flight in 1948; (d) development of Terrier, Talos, and Tartar guided missiles operational on 72 fleet ships; (e) development of the Typhon weapons system; radar, fire control equipment, computer and missiles; (f) development and implementation of concepts for operational evaluation of Fleet Ballistic Missile System; (g) development of Navy Navigational Satellite System in operational use; (h) first successful theory of the causes of unstable burning in solid rockets; (i) demonstration of thrust and lift of external burning supersonic ramjet; (j) successful three axis gravity-gradient stabilized satellite at near-synchronous altitude; (k) first unambiguous radar detection of clear air turbulence.

(11) STAFF COMPOSITION:

115	Ph.D.	135	Physicists
257	M.S., M.A.	758	Engineers and Engrg. Staff Associates
714	Other Professional	132	Mathematicians
781	Non-Prof. & Tech.	34	Chemists
699	Admin. & Clerical	27	Other
2566	Total Employment	1086	Total Professional Scientists and Engineers

(12) UNIVERSITY RELATIONSHIPS: APL is legally a Division of The Johns Hopkins University and oversight is maintained of its activities by a standing committee of the Board of Trustees. APL staff members do not have tenure. Since 1958 staff members have taught evening courses at APL, and since 1963 the Howard County Laboratory has been a Center within the JHU Evening College. Current courses are being offered to 442 students, two-thirds of whom come from 89 different organizations, and 265 of whom are M.S. candidates in four degree programs for which all courses are given at APL. To bring to bear on medical technology the fruits of Defense and NASA research and engineering, there has been an expanding program of cooperation with the Johns Hopkins Medical Institutions. Johns Hopkins is uniquely suited to lead in this effort to achieve a union of modern technology and medicine because it has in its family outstanding medical institutions and the Applied Physics Laboratory. Since 1956, two APL staff members annually receive fellowships for advanced study or teaching at JHU.

(13) ADDITIONAL COMMENTS:

Detailed References: (a) U.S. House of Representatives, Committee on Government Operations, Military Operations Subcommittee, Hearings, Systems Development and Management, (1962, pp. 1629-1642). (b) "The Johns Hopkins University Applied Physics Laboratory, A Short Account of its Environment and Work," (Rev. April 29, 1966), APL/JHU internal document.

From 1946-48, the Kellex Corp. (later the Vitro Corp.) conducted joint operations with APL-JHU at Silver Spring, Md. Since then Vitro has operated as a Navy-sponsored contractor and later under multi-agency sponsorship. Budget figures in item (7) are contract expenditures, including subcontract expenditures.

* Nominal.

** Includes affiliated staff on Kellex Corp. assigned to APL.

App. 7-1 — Data, Applied Physics Laboratory (Continued)

DATE: November 29, 1967

(1) NAME: Jet Propulsion Laboratory (JPL)

(2) LOCATION(S): 4800 Oak Grove Drive, Pasadena, California 91103

(3) SPONSORING AGENCY: National Aeronautics and Space Administration (NASA) since 1958; DOD prior to 1958
(4) MANAGEMENT ORGANIZATION: California Institute of Technology (Caltech)

(5) CHIEF ADMINISTRATOR: Dr. William H. Pickering, Director
(6) CUSTOMERS SERVED: NASA (95.5%), Other (0.5%)

(7) TABLE OF BUDGET AND PERSONNEL:

			CY	CY	CY	CY	CY						
Year (FY ending 6/30)	1936	1940	1942	1943	1946	1947	1948	1949	1950	1951	1952	1953	1954
Total Budget ($ Million)			.2	.3	2.4	3.0	3.0	2.6	2.9	5.6	10.6	12.2	9.4
R & D Budget ($ Million)													
Total Personnel									587	742	1021	1073	1106

(Continued)

Year (FY ending 6/30)	1955	1956	1957	1958	1959	1960	1961	1962	1963	1964	1965	1966	1967
Total Budget ($ Million)	11.2	17.7	26.8	37.0	45.9	62.9	78.9	139.7	183.5	249.3	234.6	244.5	229.3
R & D Budget ($ Million)					45.8	59.5	65.4	130.3	172.8	219.8	221.9	240.6	226.3
Total Personnel	1213	1495	1917	2278	2680	2770	2847	3539	4045	4324	4070	4215	4586

(8) ORIGIN OF CENTER: In 1936, Dr. Theodore von Karman, Director of the Guggenheim Aeronautical Laboratory at Caltech, encouraged a group of scientists and students to work with him in conducting experiments with rocket engines. The initial funding for these experiments was arranged through private donors, and the experiments were carried out in a then desolate spot in Arroyo Seco, north of Pasadena, on which site the Laboratory now stands. It was not until 1939 that significant funding came from the military and the National Academy of Science to assist in the development of rockets to help Army Air Corps planes at take-off.

(9) ORIGINAL AND SUBSEQUENT MISSIONS: Between 1939 and December, 1958, JPL performed R & D in solid propellant fuels, tactical missiles, and rocketry for the Department of the Army. The newly formed National Aeronautics and Space Administration took over responsibility for JPL in 1958 from the Army and has since broadened the Laboratory's mission to extend from the most basic research tasks to the most complex development activities of automated space vehicles and operation and maintenance of worldwide tracking and data acquisition networks. In 1967, flight hardware development and project management constituted JPL's largest effort. NASA has delegated the responsibility to JPL for the development of spacecraft including the Ranger, Surveyor, Mariner II, IV and V, Mariner '67 (Venus mission), Mariner '69 (Mars mission), and the development of the overall plan and technical direction of portions of the Voyager program which NASA has proposed to commence missions to Mars in 1973-75. JPL functions in many ways similar to some NASA in-house laboratories. NASA unilaterally issues task orders directing JPL to perform specific tasks in the areas of flight projects, deep space instrumentation, and supporting research and technology. JPL is also given complete responsibility for major spacecraft programs including contracting with industrial firms and supervising the work of these firms.

App. 7-2 — Summary of important data on Jet Propulsion Laboratory, a major contract R&D center for NASA.

(10) MAJOR ACCOMPLISHMENTS: During the period from 1939-1950 JPL pioneered the development of both liquid and solid fuel rocket motors in the United States. In 1946 invented case bonded solid propellant system. 1944-46 developed FM-FM telemetry systems for use with guided missiles. JPL developed the United States' first surface-to-surface guided missile system "Corporal" and later the all inertial solid propellant surface-to-surface missile system "Sergeant," 1944-1960. Jan. 31, 1958, JPL with the Army team at Huntsville, launched the United States' first satellite, Explorer I. Beginning about 1958, JPL developed low-noise phase-coherent deep space communications and tracking systems. JPL designed and successfully flew the first United States spacecraft to the moon and to the planets which provided high resolution TV pictures of the moon and planets via Ranger, Surveyor, and Mariner from 1964 to the present. JPL managed the Surveyor project which accomplished the first soft-landing on the moon and gave detailed information on the lunar surface composition and physical properties (1961-1968).

(11) STAFF COMPOSITION:

1732	Scientists and Engineers	193	Ph. D.
1139	Administrative	473	M. S.
819	Clerical	868	B. S.
135	Service	1337	Other Professional (Non degree)
761	Technician	2871	Total Professional Staff
4586	Total Employment		

(12) UNIVERSITY RELATIONSHIPS: JPL is operated as a division of the California Institute of Technology. Dr. W. H. Pickering, Director, is a member of the faculty and is responsible to the President for the operation of the Laboratory. In FY 1967, 14 Caltech faculty members served part time at JPL. In the summer of 1967, 18 Caltech graduate students were employed at JPL. Fourteen (14) faculty members from universities other than Caltech were employed during 1967. Approximately 20 scientists from 20 universities have served as principal investigators on JPL-managed NASA space flight projects. There are two standing committees at Caltech concerned exclusively with JPL: A faculty committee for scientific and technical matters, and a committee of the Board of Trustees concerned with management and the overall policy of JPL. In addition, there are approximately 12 faculty members who are paid consultants to JPL in various scientific disciplines and 16 research associates on its staff representing five foreign countries.

(13) ADDITIONAL COMMENTS: The Government-owned plant and equipment used by JPL is valued at over $139 million.

Detailed References: (a) Jet Propulsion Laboratory, 1965 Annual Report, 48 p., 1966 Annual Report, 40 p.; (b) U.S. National Aeronautics and Space Administration, An Administrative History of NASA, 1958-1963 (Washington: U.S. Government Printing Office, 1966), pp. 45-50, 54-55, 59, 74-75, 93-95, 155, 174, 292; (c) U.S. Congress, House of Representatives, Committee on Government Operations, Military Operations Subcommittee, Systems Development and Management, Hearings, June 21-August 21, 1962 (87th Congress, 2nd Session) (Washington: U.S. Government Printing Office, 1962, pp. 1735-1750).

App. 7-2 **— Data, Jet Propulsion Laboratory (Continued)**

(1) NAME: Oak Ridge National Laboratory (ORNL) (Clinton Laboratories, 1943-March 1, 1948) **DATE: October 30, 1967**

(2) LOCATION(S): Oak Ridge, Tennessee 37830

(3) SPONSORING AGENCY: U. S. Atomic Energy Commission

(4) MANAGEMENT ORGANIZATION: Union Carbide Nuclear Company, Division of Union Carbide Corporation

(5) CHIEF ADMINISTRATOR: Alvin M. Weinberg, Director

(6) CUSTOMERS SERVED: FY 1967 - USAEC (88. 9%), Army (1. 0%), DASA (0. 7%), NASA (2. 1%), NIH (3. 9%), OSW (1. 4%), ARPA (0. 2%), NBS (0. 1%), Air Force (-%), PHS (-%), AID (-%), USFS (-%), and Non-Governmental Clients (1. 6%).

(7) TABLE OF BUDGET AND PERSONNEL:

Year (FY ending 6/30)	1949	1950	1951	1952	1953	1954	1955	1956	1957	1958	1959	1960	1961
Total Budget ($ Million)	14. 1	15. 5	27. 9	34. 8	36. 7	37. 2	41. 2	51. 4	59. 6	58. 4	64. 4	67. 5	77. 3
R & D Budget ($ Million)	10. 5	11. 4	19. 6	26. 2	28. 6	28. 9	32. 8	42. 4	51. 3	47. 0	51. 1	54. 6	61. 1
Total Personnel	2034	2340	3025	3246	3240	3436	3519	3798	3947	4189	4396	4172	4124

(Continued)

Year (FY ending 6/30)	1962	1963	1964	1965	1966	1967							
Total Budget ($ Million)	81. 2	89. 7	90. 2	98. 4	106. 9	111. 1							
R & D Budget ($ Million)	62. 1	65. 2	66. 6	69. 5	74. 6	79. 2							
Total Personnel	4251	4458	4666	4824	4942	5006							

(8) ORIGIN OF CENTER: Oak Ridge National Laboratory, successor of the wartime Clinton Laboratories, was organized around the world's first operating reactor (completed in 11 months after E. Fermi's experiment on the self-sustained nuclear chain reaction) to serve as a pilot plant for Hanford Works' large production reactors and to develop plutonium-extraction methods. The first contractor, the University of Chicago collaborated closely with the DuPont Company, and was succeeded briefly by the Monsanto Company. In 1948 operation of the facility was taken over by Carbide and Carbon Chemicals Company, a unit of Union Carbide Corporation, which is responsible for the other two large AEC projects. At that time the Laboratory assumed its present name. At present, the Laboratory also occupies several large buildings in the Y-12 Plant (where ^{235}U was separated by the electromagnetic process during the war) and expanded into another valley from its original location, known as the X-10 Site.

(9) ORIGINAL AND SUBSEQUENT MISSIONS: The original research programs centered on reactor technology and chemical extraction methods. The primary goal was to demonstrate the feasibility of the large-scale production of plutonium in a nuclear reactor and to select the best configuration for the reactor itself; thus, problems of the radiolysis of the water used to cool the Hanford reactors and plutonium extraction techniques were examined. The new interdisciplinary science of health physics and supporting research programs in physics and chemistry received special attention. After WWII, a balanced applied and fundamental research program was established. The main applied areas in 1967 are reactor technology ($11. 0 million), civilian power reactors ($13. 4 million), nuclear safety ($6. 1 million), radioactive and stable isotope research and development ($5. 6 million). The major fundamental research areas include chemistry ($13. 4 million), biomedical sciences ($11. 0 million) and low energy physics ($3. 1 million). Study of new methods of information handling is also actively pursued.

App. 7-3 — Data summary on Oak Ridge National Laboratory, a major, multidiscipline R&D center operated for AEC by Union Carbide Corp.

At present, ORNL is a multidisciplinary institution with diversified AEC programs and projects sponsored by other agencies such as the Office of Saline Water (Department of Interior), National Institutes of Health, Public Health Service (Department of Health, Education and Welfare), Defense Atomic Support Agency, Office of Civil Defense (Department of Defense), etc. Oak Ridge is engaged in basic research (40%), applied research (20%), and development (40%).

(10) MAJOR ACCOMPLISHMENTS: 1. Reactor Development: (a) Design and construction of OGR, world's first continously operating reactor; (b) development of BSR and ORR, prototypes of inexpensive, versatile, pool-type research reactors; (c) design and construction of TSF, used for shielding studies; (d) operation of ARE, molten-fluoride-fueled power reactor prototype; (e) design and operation of MSRE, demonstrating the feasibility of a molten-fluoride-fueled thermal breeder for power generation; (f) development and design of HFIR for producing transplutonium elements; (g) design of the MTR, the first very-high-flux test reactor; (h) design of APPR, a mobile reactor prototype; (i) development of the nuclear desalination concept, based on the production of fresh water and electricity by means of large reactors; (j) operation of the HPRR, producing a fast burst of $\sim 10^{18}$ n/cm^2. 2. Reactor Fuels and Materials: Developments of (a) aluminum-plate "sandwich" type fuel elements for pool reactors; (b) boiling-potassium heat-transfer technology; (c) design of stainless-steel dispersion fuel elements; (d) development of INOR-8 corrosion-resistant alloy for containing molten-fluoride mixtures; (e) the Sol-Gel process for preparing uranium, thorium and plutonium fuels; (f) study of metal fluoride phase diagrams and development of mixed, molten-fluoride fuels for power reactors. 3. Chemical Technology: (a) Development of the first plutonium-separation method by the bismuth phosphate process; (b) recovery of uranium and thorium from ores by the Dapex and Amex processes; (c) recovery of plutonium and uranium, ^{233}U-thorium and ^{235}U from irradiated fuels by the Purex, Thorex, and TBP processes, respectively; (d) treatment of radioactive liquid waste, -pot calcination process for the high-level waste liquors, and hydrofractionation for low- and intermediate-level liquids; (e) separation of actinide elements from fission products (Tramex process). 4. Isotopes: First production of a radiosotope (^{14}C) by reactor supplied nearly 100 radioisotopes (about 10 million curies) and over 250 stable isotopes to users all over the world. 5. Life Sciences: (a) Clarification of the somatic and genetic effects of radiation; (b) estimation of acceptable working levels for long-term exposure to about 250 radioisotopes; (c) discovery of transfer RNA; (d) development of ion-exchange methods for nucleotide separation; (e) development and application of the zonal-gradient liquid ultracentrifuge for biological separations. 6. Physical Sciences: (a) Development and application of shielding theory; (b) theory of radiation effects on metals and semiconductors; (c) development of silicon-diode detectors; (d) determination of η^* from ^{233}U; (e) development and application of neutron diffraction techniques to problems of molecular structure; (f) design and development of the isochronous cyclotron; (g) activation-analysis techniques.

Abbreviations of Reactors: OGR - Oak Ridge Graphite Reactor; BSR - Bulk Shielding Reactor; ORR - ORNL Research Reactor; TSF - Tower Shielding Facility; ARE - Aircraft Reactor Experiment; MSRE - Molten Salt Reactor Experiment; HFIR - High-Flux Isotopic Reactor; MTR - Materials Testing Reactor; APPR - Army Package Power Reactor; HPRR - Health Physics Research Reactor.

App. 7-3 — Data, Oak Ridge National Laboratory (Continued)

(11) STAFF COMPOSITION:

326 Physicists	351 Life Scientists (Bio., H.P.)	546 Ph.D.	1050 Non-Prof. & Tech.
404 Chemists	77 Other	394 M.S.	693 Admin. & Clerical
65 Mathematicians	1989 Total Professional Scientists and Engineers	1049 Other Professional	3732 Total Employment**
766 Engineers			

(12) UNIVERSITY RELATIONSHIPS: In view of its extensive education activities, ORNL has an Office of Education and University Relations which coordinates the summer employment of approximately 75 professors and more than 200 graduate and undergraduate students; forty co-op students are employed for six months per year. ORNL collaborates closely with ORAU, providing assignments for students supported by AEC fellowships for postdoctoral reasearch, thesis research, and special training in health physics. There are three schools at the Laboratory: The M.I.T. Practice School, directed by 2 full-time staff members, providing eight-week periods of practical operating and design experience for 40 chemical engineering students; the Graduate School of Biomedical Science, conducted in cooperation with U.T. with a present complement of six students and a full-time staff of 3; the U.T. Practice School for nuclear engineering students to be started soon. In addition to the 45 part-time professors on the ORNL staff at U.T. on a regular basis, a number of others teach evening courses outside Laboratory working hours.

Abbreviations: ORAU - Oak Ridge Associated Universities, a joint agency of 42 universites. U.T. - University of Tennessee.

(13) ADDITIONAL COMMENTS:

* eta -- neutron regeneration factor
** Plus 1264 Bargaining Unit Employees

App. 7-3 — Data, Oak Ridge National Laboratory (Continued)

APPENDIX 9-1

MICROPROCESSOR-BASED SATELLITE-CONTROLLED CLOCK

D. W. Hanson, D. D. Davis, and J. V. Cateora

17623

National Bureau of Standards, Boulder, Colorado

Abstract

As a demonstration of the usefulness of the NBS-sponsored time signals from two U.S. weather satellites, NBS designed and built a "Satellite-Controlled Clock." The initial design used random logic requiring nearly 80 packages of TTL logic. With the advent of microprocessors, the clock was redesigned using a four bit microprocessor and a scientific calculator chip.

Today, commerical versions of the NBS-designed clock are available and are finding use in the electric power industry, radio and television broadcasting, defense, communications, and geophysical monitoring.

The evolution of the satellite-controlled clock from random logic to commercial products using eight bit microprocessors are discussed in some detail.

Introduction

In May 1974, the Time and Frequency Division of the National Bureau of Standards (NBS) added a time code to a continuous data stream transmitted from the National Oceanic and Atmospheric Administration's (NOAA's) Geostationary Operational Environmental Satellite (GOES) satellites. The time code not only contained the usual information of days, hours, minutes, seconds and a UT1 (astronomical time scale) correction but added new information. That information was included to provide the user the capability of computing the propagation delays between the master clock and his location via the satellite. That information was the transmitting satellite's current position in terms of it's subsatellite longitude, latitude, and height above the surface of the earth.

The combination of a satellite-disseminated time code and an easily computed propagation delay had all the elements of a greatly advanced system for time distribution. To demonstrate the capabilities of this advanced timing system and to accelerate and encourage the transfer of its technology to users and industry, NBS developed the instrumentation, a decoder clock, to recognize the time code, set, and control ground based clocks. This instrumentation was first embodied in mid 1974 as random logic circuitry. The advent of microprocessors led NBS in 1975 to replace the random logic design with a 4004* microprocessor yielding an improved instrument which was completely automatic.

Today, this work has fostered the development of similar commercial instruments using 8080 and 6800 microprocessors. These instruments are now being used in the electric power industry, defense, geophysics, communications and the broadcast industry.

The evolution of the satellite-controlled clock is discussed in this paper and adds to the growing list of illustrations of microprocessors simplifying what was once a complicated if not impossible instrumentation problem.

Details including schematics, board layouts, flowcharts, and software of the actual NBS designed instruments have been completely documented in the references and therefore will not be discussed in this paper.

Definition of the Instrumentation Problem

Available from a receiver (the receiver is not part of the instrumentation development to be described here) are two outputs--data and data clock. The data consists of binary "0's" and "1's" each 10 ms in duration. The data clock is a 100 Hz square wave with its negative going transitions simultaneous with the mid point of data bits (see figure 1).

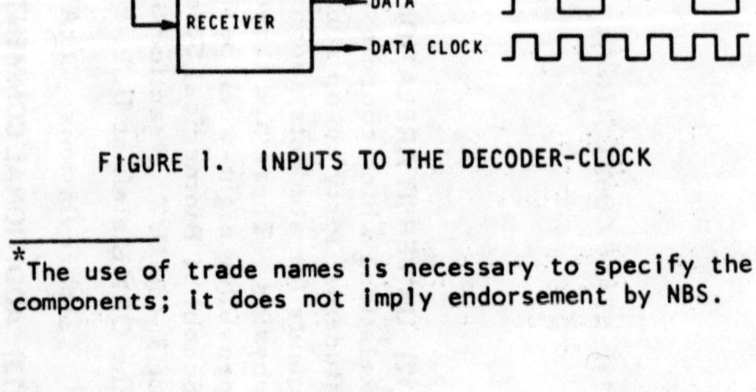

FIGURE 1. INPUTS TO THE DECODER-CLOCK

*The use of trade names is necessary to specify the components; it does not imply endorsement by NBS.

The data stream contains the time code which is time division multiplexed with other data. The data stream consists of repeated sequences of frames of four bits representing a binary coded decimal (BCD) word of the time code beginning on the one-half second followed by a maximum length sequence (MLS) 15 bits in length used for message synchronization and ending with 31 bits of data not related to the time code or frame synchronization (see figure 2).

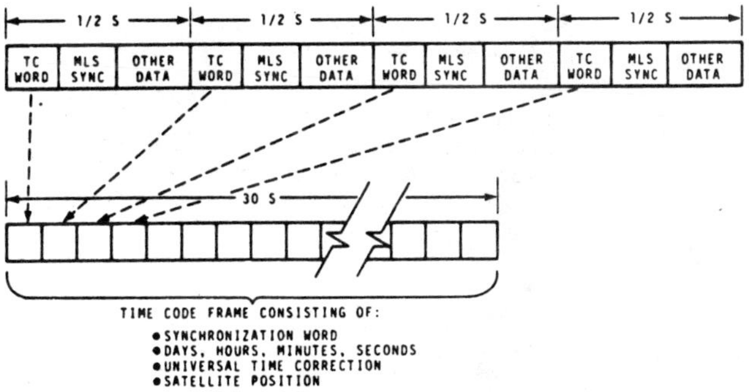

FIGURE 2. GOES DATA STREAM

The data rate is 100 b/s and is determined by atomic oscillators. Each message frame is one-half second in length (50 bits). The time-code frame, made up by gathering together four bits from each frame begins on the half minute as shown in figure 3.

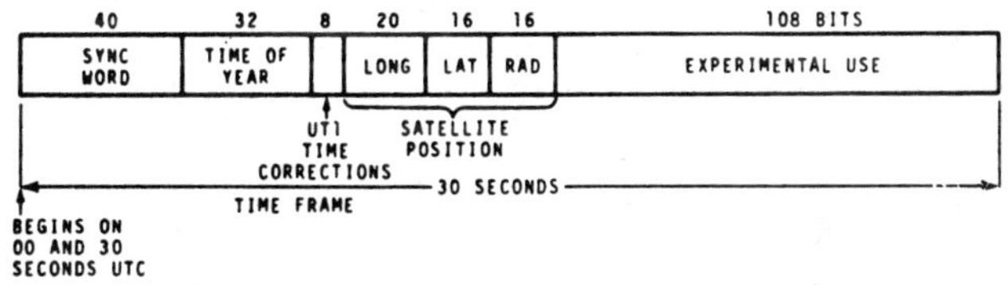

FIGURE 3. TIME-CODE FORMAT

The time-code frame contains its own synchronization word, a time-of-year word, UT1 correction, and the satellite's position.

A decoder clock, to be effective, must perform, as a minimum, the following functions:

1. Detect the negative going transitions of the data clock and immediately sample the data to determine if it is a binary "0" or "1".

2. Search 15 consecutive bits of data for the MLS synchronization pattern (100010011010111).

3. From the end of the frame sync word, "count" 31 bits to the beginning of a time-code word, extract four bits, and store.

4. Examine ten consecutive time-code words for the time-code synchronization word and detect its occurrence.

5. Set the internal "date clock" with the received time-of-year word, the location of which was found in step 4. above.

6. "Store" UT1 and satellite position data.

7. Update the date clock by counting data-clock transitions; each count incrementing the date clock by 10 ms.

8. Compare the date clock with subsequent time-of-year messages received from the satellite.

9. Display decoded data and the time of year.

10. Output a 1 pps electrical pulse and a standard time code such as IRIG-B.

Instrumentation

Random Logic Solution

The above ten functions were first realized in 1974 using entirely transistor transistor logic (TTL) random logic. Figure 4 shows the instrument which contained nearly 80 integrated circuit packages. The unit presented significant cost, power consumption, and reliability problems that were undesirable and prompted NBS to look for simpler solutions.

FIGURE 4. RANDOM LOGIC DECODER CLOCK

This decoder clock also had no computational capabilities. NBS compensated for this deficiency by designing a plastic slide rule to compute the propagation delays. The slide rule, shown in figure 5, computed the free-space distance, r, in microseconds between the ground and satellite. The distance, r, in microseconds is given by

$$r = \frac{\sqrt{R^2 + h^2 - 2Rh\cos\beta},}{C}$$

where C = 0.2997925 km/μs,

$$h = a\sqrt{\frac{1 + \frac{b^4}{a^4}\tan^2\phi}{1 + \frac{b^2}{a^2}\tan^2\phi}},$$

$$\cos\beta = \sin\phi_r \sin\phi_s + \cos\phi'_r \cos\phi_s \cos|\lambda_s - \lambda_r|,$$

$$\tan\phi' = \frac{b^2}{a^2}\tan\phi,$$

r is the distance from the earth's center to the satellite,

ϕ' is geocentric latitude,

ϕ is geodetic latitude,

b is the earth's semiminor axis (6356.5838 km),

a is the earth's semimajor axis (6378.2064 km),

and the subscripts r and s refer to the earth's surface and satellite respectively.

FIGURE 5. PROPAGATION DELAY SLIDE RULE

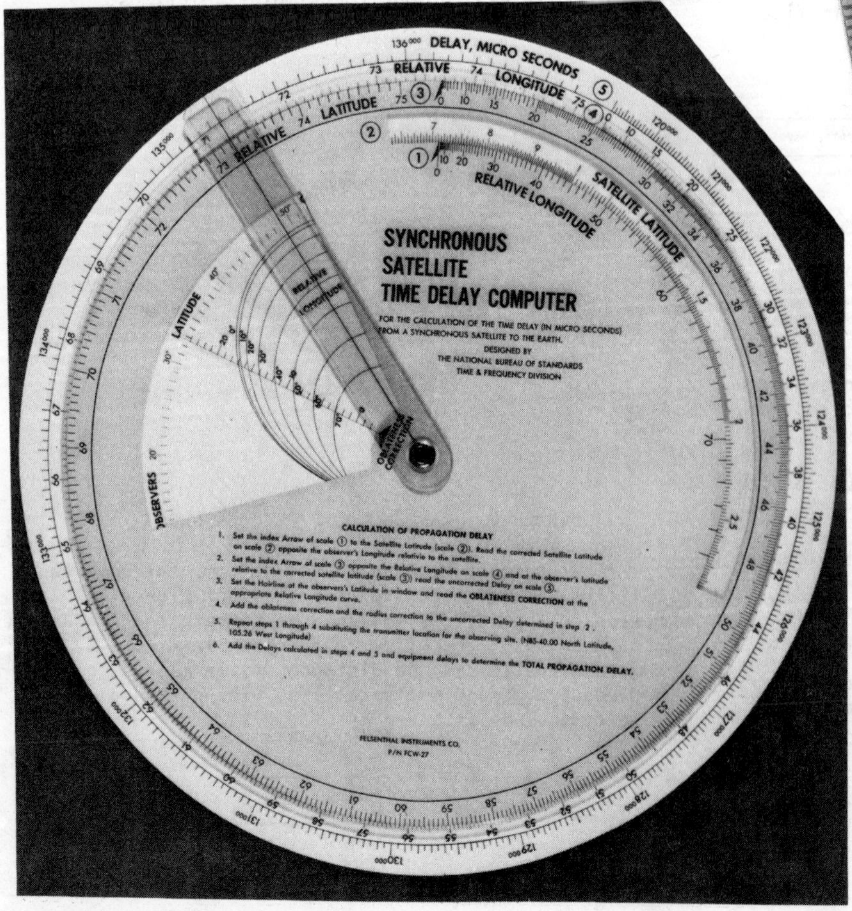

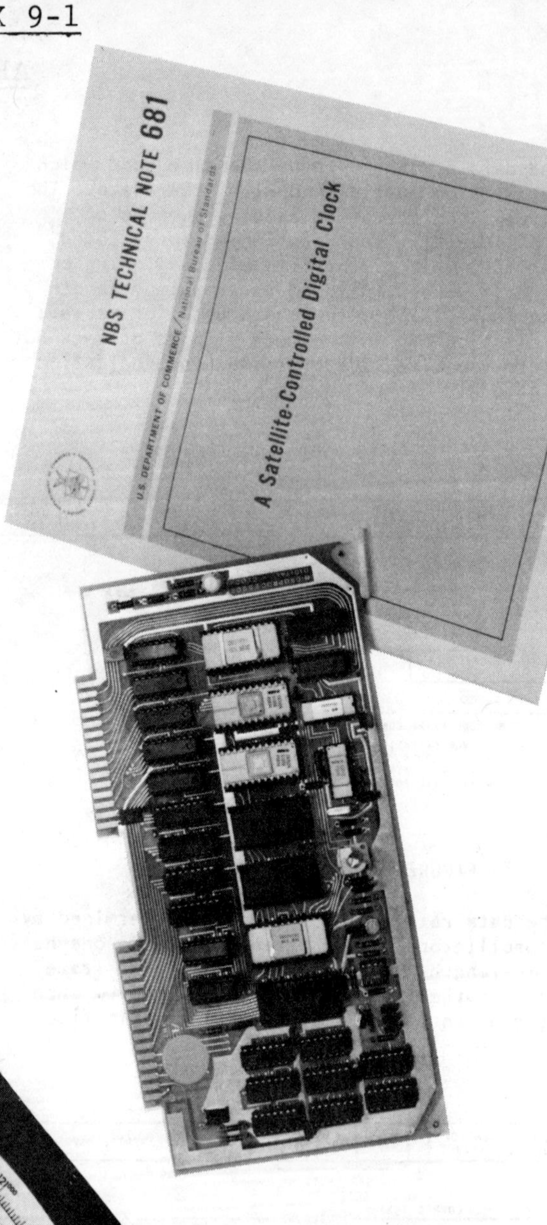

FIGURE 6. SATELLITE-CONTROLLED CLOCK USING THE 4004 MICROPROCESSOR

Four-bit Microprocessor Solution

It was fortuitous that microprocessors were becomming available when NBS began looking for a means of simplifying and improving the random-logic based decoder clock. NBS accomplished a new instrument design using a 4004 microprocessor and integrating it with a scientific calculator chip to perform the path delay calculations discussed previously. This microprocessor-based instrument is shown in figure 6. The unit, in addition to the ten basic functions mentioned, computed delays between the master clock and its location on the earth's surface via the satellite. The block diagram in figure 7 illustrates the functions. Thumb wheel switches input the user's longitude and latitude. The microprocessor and calculator chip computes delays and compensates for them by controlling a programmable delay generator so the 1 pps output is in synchronism with the NBS master clock Coordinated Universal Time (UTC)NBS.

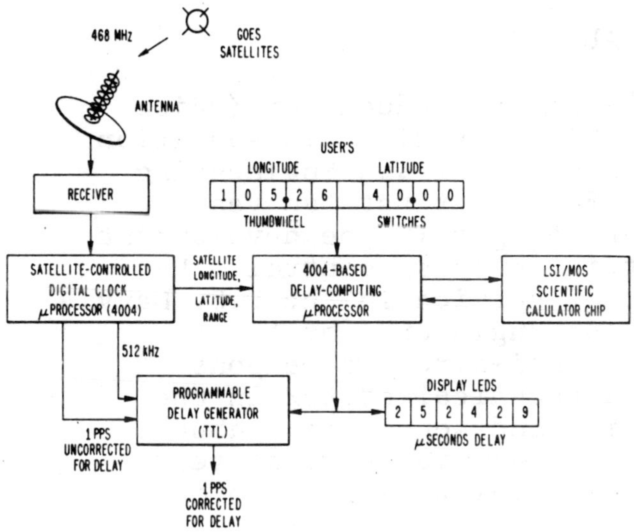

FIGURE 7. BLOCK DIAGRAM OF 4004 DECODER CLOCK

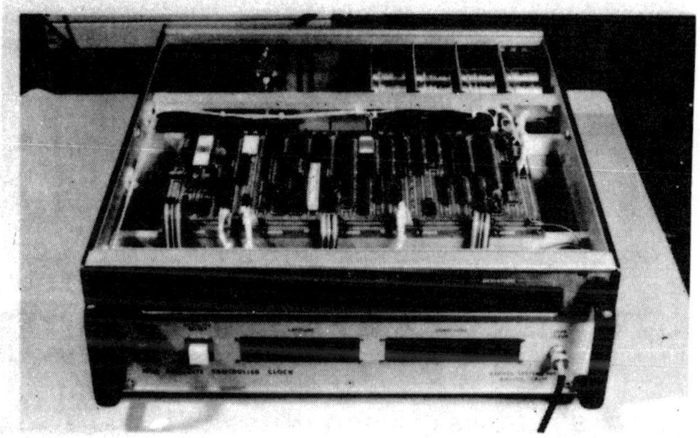

FIGURE 8. COMMERCIAL "SMART" CLOCK USING AN 8080

FIGURE 9. COMMERCIAL CLOCK USING AN 6800

Commercial Clocks

The basic approach to the satellite-controlled clock taken by NBS has recently led to two commercial versions of the instrument. One unit is the "smart" clock shown in figure 8 and uses a 8080 microprocessor. It sells for a little over $4000 including a receiver and antenna. Another version uses a 6800 microprocessor to obtain all functions except automatic path delay corrections. This clock and receiver, shown in figure 9, sells for approximately $2000 depending upon options.

NBS has been contacted by more than 18 manufacturers interested in developing products of a similar nature.

Conclusions

The capability of the microprocessor to simplify and add capabilities to the satellite-controlled clock was shown. The satellite-controlled clock, by virtue of using satellite-transmitted signals and being completely automatic by use of microprocessors, has made the GOES time dissemination system the most viable and exciting new source for synchronization today. This is particularly true for users with more than casual interests. Presently, the GOES time system is being considered for use with European and Japanese satellites. A major reason for its acceptance by other countries has been its automatic recovery features made possible by microprocessors.

References

1. Cateora, J. V., Davis, D. D. and Hanson, D. W., "A Satellite-Controlled Digital Clock," NBS Technical Note 681 (June 1976).

2. Hamilton, W. F. and Hanson, D. W., "A Synchronous Satellite Time Delay Computer," NBS Technical Note 638 (July 1973).

3. Cateora, J. V., Hanson, D. W. and Davis, D. D., "Automatic Path Delay Corrections to GOES Satellite Time Broadcasts," NBS Technical Note 1003 (to be published April 1978).

ABOUT THE AUTHOR

Hyman Olken acquired a liberal arts education (major in history) at Harvard College while pursuing his education in electronics (radio) engineering at Harvard Engineering School (B.A. 1929, M.A. 1930). For his first ten years after college he was an engineer engaged in the development of automatic devices for industrial process control. He next had a long (15 years) span of service in the U.S. Navy Department, first as a development engineer (for three years) on automatic controls for anti-aircraft weapons systems, then several years as a technical editor in the production of equipment (operation and servicing) manuals on these weapons systems. Then he was, for several years, editor in the production of Navy training manuals on ordnance and electronics.

This was followed by two years of service in the first U.S. Government agency for the promotion of technology transfer--the Office of Technical Services (now the National Technical Information Service). He next worked for three years in an advertising agency where he learned the marketing of industrial products.

Mr. Olken then began a 16-year span of service in the Electronics Engineering Department of the Lawrence Livermore (National) Laboratory in Livermore, California. Here he instituted and administered a modular system of preparing procurement specifications on high technology research equipment. This gave him a front-line view of the process of new technology development in Government R&D centers.

For the past several years Mr. Olken has been retired. He spends his retirement in writing on technology transfer. Besides this book, his writings on the subject include the following:

1. A report to the Administrator of the State Technical Services Act for the State of California entitled INFORMATION RESOURCES FOR PROMOTING ADVANCED TECHNOLOGY DEVELOPMENT IN CALIFORNIA. A Survey of Present Resources and Suggestions for Their Utilization and Future Development. Berkeley, CA Nov. 16, 1966.

Part I; Types of Industrial Firms, Their Spin-off Needs, and the Techniques for Meeting Them, 12 pp.

Part II: Directory of Technical Information Resources, Vol. I - Advanced Technology Industries, 285 pp.

Vol. II - Old-Line Industries, 156 pp.

2. Spin-Offs - A Business Pay-Off. CALIFORNIA MANAGEMENT REVIEW, Winter issue, 1966, pp. 17-24.

3. Spin-Offs II - Why Has Industry Made So Little Use of Government-Created Technological Advances? CALIFORNIA MANAGEMENT REVIEW, Winter Issue, 1967, pp. 35-42.

4. TECHNOLOGY TRANSFER: How To Make It Work. Olken Publications, 2830 Kennedy Street, Livermore, CA 94550. 1972, 96 pp.

Other Books

By

This Author

TECHNOLOGY TRANSFER:

HOW TO MAKE IT WORK

Olken Publications, 1972

ISBN 0-934818-00-2